ROUTLEDGE LIBRAH
URBAN PLAN

Volume 13

NEIGHBORHOOD JOBS, RACE, AND SKILLS

NEIGHBORHOOD JOBS, RACE, AND SKILLS

Urban Unemployment and Commuting

DANIEL IMMERGLUCK

Routledge
Taylor & Francis Group

LONDON AND NEW YORK

First published in 1998 by Garland Publishing, Inc.

This edition first published in 2018
by Routledge
2 Park Square, Milton Park, Abingdon, Oxon OX14 4RN

and by Routledge
711 Third Avenue, New York, NY 10017

Routledge is an imprint of the Taylor & Francis Group, an informa business

© 1998 Daniel Immergluck

All rights reserved. No part of this book may be reprinted or reproduced or utilised in any form or by any electronic, mechanical, or other means, now known or hereafter invented, including photocopying and recording, or in any information storage or retrieval system, without permission in writing from the publishers.

Trademark notice: Product or corporate names may be trademarks or registered trademarks, and are used only for identification and explanation without intent to infringe.

British Library Cataloguing in Publication Data
A catalogue record for this book is available from the British Library

ISBN: 978-1-138-49611-8 (Set)
ISBN: 978-1-351-02214-9 (Set) (ebk)
ISBN: 978-1-138-48625-6 (Volume 13) (hbk)
ISBN: 978-1-138-48627-0 (Volume 13) (pbk)
ISBN: 978-1-351-04595-7 (Volume 13) (ebk)

Publisher's Note
The publisher has gone to great lengths to ensure the quality of this reprint but points out that some imperfections in the original copies may be apparent.

Disclaimer
The publisher has made every effort to trace copyright holders and would welcome correspondence from those they have been unable to trace.

NEIGHBORHOOD JOBS, RACE, AND SKILLS

URBAN UNEMPLOYMENT AND COMMUTING

———————————

DANIEL IMMERGLUCK

GARLAND PUBLISHING, INC.
A MEMBER OF THE TAYLOR & FRANCIS GROUP
NEW YORK & LONDON / 1998

Copyright © 1998 Daniel Immergluck
All rights reserved

Library of Congress Cataloging-in-Publication Data

Immergluck, Daniel.
Neighborhood jobs, race, and skills : urban unemployment
and commuting / Daniel Immergluck.
p.   cm. — (Garland studies in the history of American
labor)
Updated and revised version of the author's Ph. D. disserta-
tion.
Includes bibliographical references and index.
ISBN 0-8153-3207-6 (alk. paper)
1. Urban poor—Employment—Government policy—United
States. 2. Hard-core unemployed—Government policy—United
States. 3. Industrial location—Government policy—United States.
4. Enterprise zones—United States. 5. Job creation—United
States. 6. Neighborhood—United States. 7. Inner cities—United
States I. Title. II. Series.
HD5708.85.U6I45 1998
331.13'7973—dc21

                                                    98-28605

Printed on acid-free, 250-year-life paper
Manufactured in the United States of America

Contents

Preface

Over the last thirty years, a good deal of research has been focused on the employment prospects of urban residents, particularly minorities living in the inner city. Various scholars, elected officials, and planners have addressed urban employment problems by calling for targeting labor demand or supply at the national, state, local and neighborhood-levels. Positive labor demand shocks at the metropolitan level are shown to improve the employment and, particularly, the earnings prospects of black males (Bartik, 1991; Freeman, 1991). This effect lies in the last-in-first-out phenomena of labor markets, where low-skill blacks are hired after the economy has expanded enough to exhaust the supply of low-skill nonblack labor. Intermetropolitan analyses, however, do not convincingly argue that metropolitan job growth will permanently and substantially reduce unemployment in neighborhoods that have suffered from large increases in unemployment over the last thirty years.

   Place-based, neighborhood-targeted job creation is often seen as a more direct demand-side attack on the problems of high neighborhood unemployment. Such policies are concerned much more with the spatial distribution of jobs within a metropolitan area and much less with aggregate job growth of an entire metropolitan area. Besides limited place-based programs, policy makers over the last three decades have attempted generally modest programs of housing and transportation mobility programs to address the spatial concentration of minority and low-income households in inner city areas, where jobs have become more scarce. The place-based proposals, some of the earliest of which Kain and Persky (1969) derided as "gilding the ghetto," have often been criticized as unfeasible, due to the significant obstacles to business development in such areas. Criticisms regarding political feasibility and scale have been leveled against housing mobility programs, in which inner-city residents are relocated into new areas, ideally more affluent ones closer to growing job centers. Finally, Hughes (1989) and others have called for reverse commuting programs to help inner-city residents find and commute to jobs in the distant locales without relocating residences. This strategy, too, has been criticized for issues of feasibility and a failure to address problems of social isolation more comprehensively.

The spatial mismatch problem has now become so well known that some policy-makers and planners, who may be only indirectly or vaguely familiar with the research, talk as if commuting distance, alone, were known to be the most important barrier to increasing the employment rates of inner-city residents. A careful reading of the literature reveals a good deal of uncertainty regarding the importance of various barriers to employment. The literature is quite voluminous and is methodologically complex. Many policy makers know that the job growth is typically occurring far from inner-city areas, and then readily accept that this phenomena is a principal cause of the poor employment prospects of inner-city residents. While it is true that job proximity has frequently been found to have some effect on employment rates, the importance of the effects vis-à-vis other factors is not entirely clear. Moreover, it may be that the spatial distribution of jobs is much more important in determining what types of jobs inner-city residents hold and how far they must commute than whether they hold a job of any sort. In addition, it may be that firms far from inner-city areas are located in such places, in large part, to avoid inner-city job applicants (Turner, 1997). Thus, a lack of local jobs may be as much a result of discrimination as it is a cause of unemployment.

In addition to the complexity of the problem and the difficulty of distilling subtleties into the policy arena, much of the empirical research is not constructed so as to answer specific questions about the likely impact of particular types of programs on different outcomes. The literature often utilizes data and techniques that advance the cause of methodological sophistication and precision, even though the questions such approaches address may not be of the most relevance in actual policy space. Resources are often devoted to continuing a debate over basic social phenomena, based on the shortcomings of previous research. Often, however, such refinements are conducted while little effort is made to inform actual planning and policy design. For example, the spatial mismatch literature often continues to focus on the question, "does job proximity affect employment rates?" Questions regarding the relative magnitude of such an effect vis-à-vis other barriers such as racial discrimination and skill mismatches are typically either subordinated or even ignored. Also, research that seeks to define the geographic scale of the effect is generally lacking. If job creation within two or three miles makes a difference, for example, this has direct implications for neighborhood economic development. Finally, too little effort has focused on the impact of space on the quality of employment attained and on the commuting burdens of inner-city residents, many of whom earn meager wages.

One area that has required more specific attention is the likely effect of neighborhood job growth on the residents of a neighborhood. Place-based economic development policies, including Empowerment Zones, business development programs, state enterprise zones, and others, tend to focus on the location or expansion of firms in and near high-unemployment areas. There may be a number of important goals for such programs, including improving access to basic goods and services, improving central-city tax bases, and others. Reducing unemployment and providing nearby employment opportunities for neighborhood residents, however, are typically among the prime objectives of neighborhood economic development policies. It is these two goals that I seek to examine. To what degree does job proximity affect neighborhood employment rates and the proportion of neighborhood residents working near the neighborhood? Also, controlling for proximity, how do the race and skills of residents affect such outcomes? How are important is the skills match between neighborhood residents and nearby jobs? These are the sorts of questions that are important to neighborhood economic development policy.

In Chapter 1, I lay out the problem of increasing neighborhood unemployment and describe the history of U.S. place-based policies that have at lease partially been aimed at economic development in low-income neighborhoods. Chapter 2 contains a review of the evidence on whether place-based economic development policies targeting submetropolitan areas have been successful in job creation or retention. In Chapter 3, I then review the theory and evidence on what factors explain the employment prospects of urban residents, with a focus on the spatial mismatch literature. Chapter 4 builds primarily on Simpson (1992) to construct a model that considers the effects of nearby jobs on the employment prospects at a neighborhood level. I examine not only effects on neighborhood employment rates but also on the propensity of residents to work near the neighborhood. Instead of looking just at the magnitude of jobs in and near a neighborhood, I examine the effects of the skills of local jobs and how well they match the skills of neighborhood residents. I also consider competition for jobs from residents who live just outside the neighborhood.

Chapter 5 presents the data set used to estimate the model developed in Chapter 4. The data are based on square, half-mile by half-mile neighborhood units called quartersections. Surrounding areas are constructed which lie within two miles of each quartersection. It is these surrounding job catchment areas that are used to measure nearby job access. Chapter 6 then provides estimations of a number of versions of the basic model unemployment and "local working," or the propensity of residents to work near the neighborhood.

Finally, Chapter 7 synthesizes the findings of Chapter 6 with the other relevant literature to draw conclusions for policy and planning. These include matching neighborhood economic development efforts to the occupational backgrounds of local residents to maximize employment effects, adopting race-conscious policies aimed at reducing discrimination and improving blacks' access to job networks, and improving job training and education for residents of high unemployment neighborhoods.

Without detailing the findings of the book here, it is worth emphasizing that, at a time when race-conscious employment programs are anathema to many policy-makers, my results suggest that race is a powerful and persistent barrier to employment. When interpreted in conjunction with the existing literature, racial discrimination in hiring against blacks appears to remain a major contributor to high unemployment rates in black neighborhoods. Moreover, the additional disadvantages that blacks face in terms of poor schools and physical and social isolation compound their employment problems. While changes in the skill requirements and the location of jobs have hurt blacks disproportionately, discrimination, whether it be statistical or pure—and it is likely some of both, remains an important contributor to concentrated urban unemployment and poverty.

This book is an updated and revised version of my Ph.D. dissertation. I am indebted to members of my dissertation committee, including Therese McGuire, Joe Persky, John McDonald, Chuck Orlebeke, and Wim Wiewel, for their help and support. I am most thankful to my wife, Lilly, for her support during this and other challenges.

Neighborhood Jobs, Race, and Skills

Urban Neighborhood Unemployment and Neighborhood Economic Development Policy

INTRODUCTION

The problems of increasing poverty and declining real incomes among residents of older urban neighborhoods in the United States have been well documented over the last thirty years. A principal contributor to urban poverty and lower incomes that has garnered a great deal of research attention over this period is urban unemployment. Since Kain's seminal work on residential location and access to jobs, research on urban unemployment has been complicated by the underclass debate, by the diverse methodologies of sociologists, geographers and economists, and by an undulating national interest in the subject (Kain, 1968; Kasarda, 1993; Massey and Denton, 1994, Wilson, 1996).

In the field of economics, employment issues have typically been the domain of labor economists, while most urban issues have fallen under the umbrella of the more recent specialty of urban economics. As Simpson suggests, the result has been that empirical work on the problems of urban unemployment has not always been built on sound theoretical foundations (Simpson, 1992). Data constraints and a general bias of viewing labor markets as entirely metropolitan in nature have led many researchers to fail to distinguish among intraurban areas or neighborhoods. While this may be acceptable in relatively small cities, some of the most severe unemployment problems have occurred in large, older cities, such as Chicago, Detroit and Philadelphia, where distances, commute times and socioeconomic barriers among neighborhoods can be substantial and where the numbers and characteristics of residents and jobs vary greatly over small areas.

Sociological approaches to the problem of urban unemployment have added a layer of richness to the importance of space, as well as race and other social factors, in analyzing urban employment. This work includes analyzing the effects of concentrated poverty, weak- and strong-tied social networks, role modeling, and other phenomena. While urban economists have tended to equate space with commute times, sociologists, in particular, have focused on the institutional barriers to employment opportunity that operate across urban space, including the mechanisms of residential and employment segregation, the intricacies of job search and placement practices, the effects of concentrated poverty on social networks, and other relevant factors.

Public policy efforts at all levels of government have attempted to address urban unemployment problems through both metropolitan and submetropolitan approaches. Research suggests that positive labor demand shocks to metropolitan economies improve the employment and, particularly, the earnings prospects of black males (Bartik, 1991; Freeman, 1991). Such analysis, however, does not convincingly argue that metropolitan job growth will permanently reduce unemployment substantially in neighborhoods that have suffered from large increases in unemployment over the last thirty years.

Neighborhood-targeted job creation policy and planning efforts are often seen as a more direct demand-side attack on the problems of chronic and growing neighborhood unemployment. These efforts are concerned much more with the spatial distribution of jobs within a metropolitan area and much less with aggregate job growth of an entire metropolitan area. Notwithstanding disagreement over the policy remedies, the continual research on Kain's spatial mismatch theory seems to corroborate distance as a barrier to employment for urban residents. Neighborhood job growth also offers other potential benefits to lower-income neighborhoods, beyond directly lowering unemployment. In particular, the number of nearby job opportunities may affect the degree to which neighborhood residents work near their neighborhood. Neighborhood-targeted economic development efforts implicitly seek to increase the number of residents attaining employment in or near the neighborhood, both to reduce unemployment in the long run and to effect other benefits. Nearby jobs are beneficial for part-time working parents, who attend to their children after school, for example. Longer commutes mean longer workdays. For youth and low-skilled workers earning meager wages, the hardship of longer commutes is not strongly outweighed by earnings. Longer commutes may lead to more tenuous labor force connections. Nearby jobs may also positively impact the development of and access to job networks among nearby firms and neighborhood residents. Firms may begin to use neighborhood networks to identify new job candidates and may work with local community-based organizations to recruit workers. Thus, in the long run, a neighborhood with a high level of local working may gain a competitive advantage in access to the nearby jobs. This could lead to lower unemployment and higher earnings, depending on the nature of the jobs.

Most recently, a significant federal policy effort, the Empowerment Zone program, has been introduced to attack urban unemployment in a targeted fashion. Some, including Nicholas Lemann, argue that neighborhood-targeted economic development approaches are doomed, while others, including Michael Porter, argue that inner-city neighborhoods offer significant latent advantages for business development and job creation (Lemann, 1994; Porter, 1995). Beyond the questions regarding the feasibility of redeveloping various types of inner-city neighborhoods, there remains the question as to whether such development is likely to benefit the residents of the targeted neighborhoods. Specifically, it remains unclear the degree to which creating jobs near high unemployment neighborhoods is likely to reduce neighborhood unemployment or generate other employment-related benefits.

The central hypothesis of this book is that increased job opportunities near a neighborhood improve the employment status of neighborhood residents. This is an assumption underlying a good deal of neighborhood economic development policy that has not been directly addressed in the research. While a good deal of the spatial mismatch literature has addressed the general issue of the impact of space and distance in determining employment prospects, the methodologies in the recent literature have not been oriented to address the impact of job access at the neighborhood level. I develop a spatially uniform and consistent approach for measuring near-the-neighborhood job access.

Measuring job opportunities is not as simple as counting all the nearby jobs. It must also incorporate the level and nature of jobs in a neighborhood context, including nearby competing labor forces and the occupations of neighborhood residents. To test the hypothesis, the study seeks to identify the factors that affect two employment status outcomes among neighborhood residents in the Chicago area: 1) neighborhood unemployment; and 2) the proportion of residents in the labor force working near their neighborhoods, or what I call the "local working rate."

The research utilizes geographically aggregated data for the Chicago metropolitan region that describe people who live and people who work in small neighborhood areas. This data set includes aggregate information on the residential originations of workers and the work destinations of residents. The data are derived from a tabulation of the 1990 decennial census by the Bureau of the Census and are labeled the Census Transportation and Planning Package (CTPP)—Chicago Urban Element.

By identifying the characteristics that tend to reduce neighborhood unemployment and increase the proportion of residents working near the neighborhood, we can understand more about the potential for neighborhood-targeted economic development efforts to make progress in these areas. The findings can also suggest some complementary policies that might be needed to increase the local employment impact of conventional development policy and practice.

This chapter describes the potential benefits of nearby jobs on urban neighborhoods, the problem of persistent and growing unemployment in low- and moderate-income neighborhoods in Chicago, and the historical and present policy context for neighborhood economic development. Chapter 2 presents the theory of targeted, place-based economic development, including major arguments for and against such policies, as well as some direct evidence on the impacts of neighborhood-level economic development policies. Following this, Chapter 3 contains a review of the existing theoretical and empirical literature regarding spatial mismatch and other intraurban employment barriers, including economic and sociological research. Models are then presented in Chapter 4 for explaining the impact of various neighborhood-level characteristics, including nearby labor demand, on neighborhood unemployment and on the proportion of the neighborhood labor force working near the neighborhood. Chapter 5 provides descriptive statistics and spatial plots of the data used for estimating the models developed in Chapter 4.

Chapter 6 gives the results of the ordinary least squares (OLS) regression used to estimate these models and provides interpretations of the results. Chapter 7 then considers implications of the findings for economic development policy and practice aimed at issues of unemployment and local working.

THE POTENTIAL EFFECTS OF NEARBY JOBS ON NEIGHBORHOODS

There are several ways in which increased demand for labor at nearby firms might benefit urban neighborhoods. First, increased labor demand from nearby establishments might be expected to decrease neighborhood unemployment, increase labor force participation, or both. In particular, these effects might be due to lower job search costs or to decreased ongoing commuting costs. Effects may be particularly large in neighborhoods with large percentages of residents in the labor force who are young, lower-skilled, employed part-time, or female. These groups may have inferior access to information networks on jobs farther from their neighborhoods or may find the costs of commuting to more distant locations prohibitive, especially if the prospective job opportunities offer low wages. Higher wage workers, and their employers, are likely to be willing to invest more in job search across a wider geographic area because the payoff for a successful search is likely to exceed search costs. Moreover, the more specific nature of high-skilled work may require a more spatially expansive search.

Second, increased demand for labor might increase the number of neighborhood residents who attain employment near the neighborhood, decreasing their commuting costs and thereby increasing their wages net of commuting costs. The influence of commuting costs on workplace-residence proximity across occupational levels is theoretically ambiguous. For higher-skilled workers, the opportunity costs of commuting are higher, but may constitute a smaller percentage of overall wages than for low-skilled workers. Residential land consumption is a central factor in many neoclassical urban economic models of residential location, with more affluent households preferring larger units of land and willing to pay for them with longer commutes. Meanwhile, low-skilled workers without automobiles and without the ability to finance automobile purchases may face excessive commuting costs relative to earnings, encouraging more workplace-residence proximity.

Third, increased demand for labor might be expected to raise wages at local establishments, as employers bid up wages. These increases might be expected to accrue to neighborhood residents disproportionately.

Finally, the presence of nearby jobs may bring benefits to neighborhood residents beyond employment gains. Establishments may provide essential goods and services, such as groceries and pharmacies, to neighborhood residents. Neighborhood firms may buy goods and services directly from self-employed neighborhood residents. The presence of nearby firms can also bring added political attention to public investment and service needs. Municipal governments may invest in infrastructure, public safety and other services to retain establishments and their associated tax base and/or employment. A less commonly recognized benefit of the presence of local jobs is the maintenance of active land use for historically industrial and commercial properties. Some commercial and industrial properties, including many in older, central-city neighborhoods, may offer substantial obstacles to reuse, especially for dissimilar uses. A substantial loss in jobs or business activity may be accompanied by a surplus of obsolete or unassailable properties, which if abandoned and left to deteriorate, can become unsightly and gathering places for illicit activities, causing detrimental effects on overall neighborhood image and quality of life. Many formerly active retail and commercial streets have seen most businesses close or leave over the last twenty years. These areas have seen often seen corresponding increases in vacant lots and buildings and, frequently, liquor stores, creating havens for drug-dealing and other illicit activity.

This study will look at only two of these effects: 1) the effect of nearby labor market demand on neighborhood unemployment; and 2) the effect on the tendency of residents to attain employment near their neighborhoods—or what I term "local working." I will leave for possible future research the effects on labor force participation, wage levels, fiscal issues, and other quality of life conditions.

THE PROBLEMS OF PERSISTENT AND GROWING URBAN UNEMPLOYMENT IN CHICAGO NEIGHBORHOODS

Unemployment rates in many central cities increased dramatically during the 1970s, due only in part to higher national and regional unemployment. In Chicago, these increases generally slowed but continued in the 1980s, despite a small decrease in regional unemployment from 6.9 to 6.7 percent. Table I illustrates the actual and regionally normalized unemployment rates for the 45 low- and moderate-income Chicago community areas from 1970 to 1990.[1] Only three community areas had unemployment rates of more than 10 percent in 1970, with none having a rate above 13.4 percent. By 1980, 31 community areas had rates exceeding 10 percent, with ten having rates exceeding 20 percent. By 1990, 35 had unemployment rates of more than 10 percent, and 13 had rates of more than 20 percent.

While national and regional unemployment rates were significantly higher in 1980 than in 1970, partly due to higher labor force participation rates, the increases cannot account for the often two- and three-fold increases in unemployment rates in many neighborhoods. This is shown by examining changes in the regionally normalized unemployment rates for the community areas from 1970 to 1980. The regionally normalized rate is simply the ratio of the neighborhood unemployment rate to the unemployment rate for the six-county metropolitan area for each census year. Many community areas saw their regionally normalized rates increase substantially. All but eight of the 45 community areas saw regionally normalized unemployment increase from 1970 to 1980. Normalized unemployment increased by more than 50 percent in ten areas, including Austin, Roseland, West Englewood, South Lawndale, South Shore, and Chatham. Most of these areas where unemployment grew substantially were predominantly black by 1980. (South Lawndale remains predominantly Hispanic.)

Table I. Raw and Regionally Normalized Unemployment Rates for the 45 Low- and Moderate-Income Chicago Community Areas

Community Area	Unemployment Rate			Normalized Unemployment Rate*		
	1970	1980	1990	1970	1980	1990
Rogers Park	2.9%	6.3%	7.6%	0.87	0.92	1.13
Uptown	4.9%	10.3%	9.5%	1.46	1.50	1.41
Lincoln Square	2.9%	6.4%	7.7%	0.87	0.93	1.15
Albany Park	2.5%	6.9%	8.5%	0.75	1.00	1.26
Hermosa	3.2%	7.8%	10.9%	0.96	1.13	1.62
Avondale	3.3%	8.3%	8.2%	0.99	1.21	1.22
Logan Square	4.9%	9.4%	10.6%	1.46	1.37	1.58
Humboldt Park	4.8%	12.9%	19.0%	1.43	1.88	2.83
West Town	6.0%	10.9%	12.6%	1.79	1.58	1.88
Austin	4.1%	13.9%	17.8%	1.22	2.02	2.65
West Garfield Park	8.0%	20.7%	26.8%	2.39	3.01	3.99
East Garfield Park	8.4%	20.6%	27.9%	2.51	2.99	4.15
Near West Side	8.0%	15.8%	20.5%	2.39	2.30	3.05
North Lawndale	8.6%	20.4%	27.3%	2.57	2.97	4.06
South Lawndale	3.9%	13.7%	13.8%	1.16	1.99	2.05
Lower West Side	6.1%	16.2%	12.6%	1.82	2.35	1.88
Near South Side	7.0%	20.2%	24.7%	2.09	2.94	3.68
Armour Square	4.9%	7.7%	12.4%	1.46	1.12	1.85
Douglas	5.4%	11.3%	17.9%	1.61	1.64	2.66
Oakland	13.4%	29.5%	45.0%	4.00	4.29	6.70
Fuller Park	11.9%	22.1%	23.9%	3.55	3.21	3.56
Grand Boulevard	9.5%	24.1%	34.1%	2.84	3.50	5.07
Kenwood	5.0%	10.0%	8.9%	1.49	1.45	1.32
Washington Park	8.0%	21.0%	31.0%	2.39	3.05	4.61
Woodlawn	7.1%	19.3%	24.2%	2.12	2.81	3.60

Table I (continued)

Community Area	Unemployment Rate			Normalized Unemployment Rate*		
	1970	1980	1990	1970	1980	1990
South Shore	4.2%	13.1%	15.6%	1.25	1.90	2.32
Chatham	3.5%	11.2%	12.6%	1.04	1.63	1.88
South Chicago	4.2%	11.4%	17.9%	1.25	1.66	2.66
Burnside	2.7%	15.2%	19.5%	0.81	2.21	2.90
Roseland	4.0%	13.5%	17.5%	1.19	1.96	2.60
Pullman	4.0%	12.9%	13.3%	1.19	1.88	1.98
South Deering	3.3%	12.7%	13.1%	0.99	1.85	1.95
East Side	2.8%	7.3%	10.6%	0.84	1.06	1.58
West Pullman	3.2%	13.4%	17.6%	0.96	1.95	2.62
Riverdale	12.1%	25.3%	34.5%	3.61	3.68	5.13
Brighton Park	3.6%	7.7%	8.7%	1.07	1.12	1.29
McKinley Park	3.7%	8.0%	8.3%	1.10	1.16	1.24
Bridgeport	4.8%	8.9%	9.2%	1.43	1.29	1.37
New City	4.8%	12.4%	18.2%	1.43	1.80	2.71
Chicago Lawn	3.5%	6.8%	11.2%	1.04	0.99	1.67
West Englewood	6.5%	20.5%	24.0%	1.94	2.98	3.57
Englewood	7.7%	18.1%	26.7%	2.30	2.63	3.97
Greater Grand Crossing	5.9%	14.1%	16.9%	1.76	2.05	2.51
Auburn Gresham	4.6%	13.1%	16.0%	1.37	1.90	2.38
Edgewater	2.6%	7.2%	7.8%	0.78	1.05	1.16
Chicago Central City	4.4%	9.8%	11.3%	1.31	1.42	1.68
6-County Metropolitan Area	3.4%	6.9%	6.7%	1.00	1.00	1.00

* Normalized by 6-County Metropolitan Area Unemployment Rate

During the 1980s, the problem of increasing unemployment in central-city Chicago neighborhoods continued. Even though regional unemployment was actually slightly lower at the time of the 1990 census than at the time of the 1980 census, unemployment increased over the decade in all but three of the 45 low- and moderate-income neighborhoods in Chicago. Eight community areas saw normalized unemployment increase by more than 50 percent.

In Chicago, as in other central cities, many of the neighborhoods hit hardest by problems of unemployment have been those that have witnessed large losses in nearby jobs, especially jobs in the manufacturing industry. While the regional unemployment rate was approximately the same in 1990 as it was in 1980, jobs in many south and west side neighborhoods declined and, simultaneously, neighborhood unemployment rates rose. From 1979 to 1989, the total number of jobs in Chicago's central city fell from 1,271,721 to 1,178,908, a relatively modest 4 percent decrease.[2] For the zip codes serving the west side of the central city, however, the total number of jobs fell 22 percent, with manufacturing employment declining by 31 percent.[3] The west side remains heavily industrial, with manufacturing constituting 37 percent of jobs in 1989, but this area's manufacturing base has declined continually. (In 1979, manufacturing jobs accounted for 41 percent of all jobs in the area.) Meanwhile, unemployment in the neighborhoods on the west side rose, collectively, from 17 percent to 21 percent from 1980 to 1990, a 25 percent increase. While zip code and Chicago community areas do not correspond, there is a great deal of overlap between the zip codes identified here and the community areas of Humboldt Park, West Town, Austin, West Garfield Park, East Garfield Park, Near West Side, North Lawndale, South Lawndale, and Lower West Side.

An important distinction among unemployment changes in west side neighborhoods is that increases in unemployment were much greater in neighborhoods with high or growing percentages of black residents, including West and East Garfield Park, the Near West Side, North Lawndale, and Humboldt Park. In West Town, South Lawndale, and the Lower West Side, areas with large and growing percentages of Hispanics, the apparent correlation to job loss was not as evident. Increases in unemployment rates in West Town and South Lawndale were only 16 percent and 1 percent over 1980 rates, respectively. The Lower West Side actually saw a 22 percent decrease in unemployment.

This suggests that characteristics other than proximity of residents to jobs may also play a significant role in employment prospects.

The loss of jobs in Chicago's central city neighborhoods and the concurrent rising unemployment rates in many of these same neighborhoods has led to policies and programs aimed at stemming the continuing loss of employers and at rebuilding neighborhood job bases. Of course, other objectives are involved in neighborhood economic development, including access to nearby retail goods and services, retention of public services, and the prevention of property abandonment and associated problems such as criminal activity. In Chicago and other cities, however, a principal objective of neighborhood economic development remains the creation and maintenance of job opportunities for neighborhood residents.

EARLY ECONOMIC DEVELOPMENT POLICIES: FEDERAL, STATE AND LOCAL POLICIES PRIOR TO WORLD WAR II

Economic development policies, as defined here, are those that are intended to alter private investment and employment decisions in specific subnational regions, states, metropolitan areas, cities or neighborhoods, with the ultimate objectives of increasing employment in the target area, diversifying and increasing the tax base, and providing local commercial services and amenities. Eisenger distinguishes economic development policies targeted at subnational areas from national economic development, which is primarily a notion of using macroeconomic policy to improve economic fortunes in the United States (Eisenger, 1988). Microeconomic, industry-specific efforts are also not considered here, except for historical context.

Eisenger differentiates traditional supply-side economic development policies from more recent demand-side policies. Supply-side policies are essentially efforts to reduce the costs or risks of doing business in a particular location through some type of subsidy. Such policies include low-interest loans, tax abatements, regulatory relaxation, and loan guarantees. Demand-side policies are efforts aimed at developing formative potential business opportunities or local capacities for business development. Such policies involve government playing a leading role in encouraging some type of activity that is expected to lead to economic growth. Demand-side policies include state venture capital programs, technology transfer and commercialization programs, and export promotion and marketing programs. Demand-side programs became much more prevalent, especially at the state level, during the 1980s.

The origins of economic development policy in the United States date back to the origin of the nation. Hansen explains that before 1789, colonial governments granted charters to companies and used tariffs to protect industries and agricultural products (Hansen, 1991). Colonies built toll roads, supported shipping and invested in waterways. While many of these investments may have been justified by a notion of providing public goods, some of this activity was directed toward supporting single firms or industries. Hansen argues that, before the Great Depression, the federal role in subnational economic development was largely indirect and did not generally constitute deliberate attempts to alter specific investment and employment decisions. The aid for infrastructure and education certainly affected local and regional growth, but aid was generally based on broad industry concerns and not targeted to specific firms or small regions. The Morrill Act of 1862 established land grant colleges and led to the proliferation of agricultural extension services. Only in a few instances was government assistance provided to encourage economic growth in a particular direction or to assist individuals. Such efforts included economic aid to individuals during Reconstruction, the Homestead Act, and land grants to railroad companies.

Eisenger suggests that, prior to World War II, although the federal government influenced subnational economic development in many ways, few federal policies were designed with economic development as a prime objective (Eisenger, 1988). He cites one clear exception to this, the intergovernmental campaign to create a national transportation network in the early part of the nineteenth century. Through land grants to states, the federal government encouraged the development of canals and roads in particular regions, primarily to encourage development. Eisenger notes that early development-oriented policies differed in two critical ways from postwar era economic development. First, such policies were episodic and did not result in sustained programs. Second, the initiative for such projects was at the state and local level. The Tennessee Valley Authority was, in many ways, an exception to general prewar federal involvement in regional economic development. This federal response to southern poverty was sustained over a long period (Hodge, 1968).

Cobb notes that local efforts to induce commercial and industrial development were occurring in the South as early as the nineteenth century (Cobb, 1993). Frequently, however, such efforts were not sanctioned by the state. A 1937 survey of 41 Tennessee communities revealed that 56 plants had received some sort of subsidy, including public bond issues approved by the state legislature in violation of the state's constitution. In 1936, the State of Mississippi instituted its Balance Agriculture with Industry (BAWI) program, the first state level economic development program. BAWI's passage represented the legitimization of public spending for economic development purposes. It authorized the use of municipal bonds for financing industrial incentives for out of state firms willing to relocate into the state.

Northern cities became actively involved in economic development dating back at least to the late nineteenth century. Sanders notes that, in large cities such as New York and Chicago, the rise of machine politics led to cities taking a stronger role in the provision of public infrastructure and the political steering of projects to foster economic development in favored districts (Sanders, 1984). Early machine politics, then, may have prompted the first examples of economic development activity at the neighborhood level.

POSTWAR ECONOMIC DEVELOPMENT POLICIES TARGETING DISTRESSED URBAN NEIGHBORHOODS

The origin of targeting development efforts to distressed urban neighborhoods or groups of neighborhoods did not reach any significant scale until after World War II. Federal development policies targeting distressed urban areas can be traced back at least to the Housing Act of 1949. The Act provided for funding local governments to cover two-thirds of the net cost of acquiring and clearing a "slum area". Local governments put up the remaining net costs of the project, either in cash or in kind. The land was then sold or leased to a private developer for a below-market price. The program, first known as urban redevelopment and later as urban renewal, was conceived primarily as a response to the federal housing shortage. At first, cities were not allowed to clear and develop predominantly commercial areas for commercial redevelopment. Through various amendments beginning in 1954, the Act allowed increasing percentages of funds to be devoted to economic development purposes, reaching 35 percent by 1965 (Eisenger, 1988).

Barnekov, et al., suggest that a key model for urban renewal was the strong private-sector guidance of the Allegheny Conference in Pittsburgh (Barnekov et al., 1989). Formed in 1943, prior to federal urban renewal efforts, the Allegheny Conference was concerned with the deterioration of downtown Pittsburgh. Forty percent of the central business district was vacant or blighted. But, according to Barnekov et al., the concern over poor economic conditions in the Pittsburgh central area was a concern not about the people who lived in or near the area, but about the difficulty in attracting desirable executives to live and work in the city. In effect, the Conference, and groups like it in other cities, represented what Cox and Mair describe as large, locally dependent firms (Cox and Mair, 1988). These included financial institutions, utilities, media, and others whose fortunes were tied in various ways to the fortunes of the city.

With the War on Poverty of the middle-1960s came a federal, and later state and local, concern with the residents of economically distressed urban neighborhoods. One of the most radical economic development programs to come out of the War on Poverty was the funding of several community development corporations (CDCs) across the country (Pierce and Steinbach, 1990). CDCs were originally envisioned as community-based nonprofit organizations that would spur investment and economic activity in a neighborhood by using federal and other funds to invest in commercial job-creating activities directly. While CDCs have been more active in developing housing in recent years, many have focused some or all of their efforts on commercial activity. CDCs work with existing private firms to revitalize neighborhoods, but they also develop, own and operate commercial ventures themselves. Advocates of CDCs argue that distressed areas are not always able to attract private investment with subsidy alone. Therefore, entirely new forms of economic establishments, with social as well as economic motives, are required (Vidal, 1995).

In 1974, Nixon's New Federalism came to urban economic development as urban renewal and Model Cities programs were replaced by a new Community Development Block Grant (CDBG). The program, remarkable for its longevity in this arena, provided a flexible source of funds for both housing and local economic development purposes. CDBG is formula funded on the basis of poverty rates, housing stock age, and other measures of distress (Eisenger, 1988). Although only a small part of CDBG funds are qualified on the basis of job creation, CDBG is a primary source of funds for economic development activities targeting low- and moderate-income urban neighborhoods, some of which include job creation as a part of their mission. These funds have been used, for example, to provide low-interest loans for businesses in targeted areas and to support CDCs.

With the election of Ronald Reagan in 1980 and a movement toward decreased federal spending on cities, a British economic development concept, the enterprise zone, was introduced in the United States The enterprise zone idea was adopted by the political right, lead by Representative Jack Kemp, but also by some urban liberals (Beaumont, 1991). British enterprise zone programs were designed to revitalize deteriorated urban areas like the Docklands that had relatively few residents (Butler, 1991). The redevelopment of vacant sites was envisioned as providing a spark to overall economic growth for a larger area. The concept was seen, in part, as a non-zero-sum policy, seeking aggregate economic growth for a region through reuse of underutilized resources.

As adopted in the United States by conservatives in the early 1980s, the enterprise zone concept was modified, so that blighted, high-crime inner-city neighborhoods were envisioned as proper targets for enterprise zones, with zone residents being the chief beneficiaries. Butler suggests that the underlying assumption of the American concept is the notion of latent potential for employment and entrepreneurship in distressed neighborhoods that are smothered by "red tape, excessive taxation, and a culture of welfare dependency" (Butler, 1991).

At the federal level, enterprise zones fared poorly during the 1980s, with only a stripped-down free trade zone concept being adopted. States picked up the enterprise zone with haste, however, as Louisiana, Connecticut, Missouri, Ohio, Pennsylvania and Rhode Island implemented some of the earliest programs. By 1989, at least 37 states had enterprise zone programs (Erickson, 1992). Beaumont suggests four reasons for rapid adoption of the concept by states (Beaumont, 1991). First, anticipated federal legislation might favor states that had programs in place already. Second, zone programs quickly became a subject of shared economic development experience across states, increasing information and training about the policy. Third, the United States Department of Housing and Urban Development (HUD) became a clearinghouse for and supporter of state programs. Finally, the overall economic development activism of state governments in the 1980s may have played a significant role.

CURRENT NEIGHBORHOOD ECONOMIC DEVELOPMENT POLICIES AND PRACTICE

Given a goal of improving the employment status of residents of high-unemployment neighborhoods, there are at least two general labor demand approaches that might be adopted. First, one can seek to increase the overall demand for labor in the metropolitan area. Bartik's work suggests that metropolitan growth disproportionately benefits those with the weakest connection to job networks and opportunities, especially black males (Bartik, 1991). Thus, to the extent that such policy is effective, state and federal policy should be aimed at increasing metropolitan growth for those cities with the greatest levels of concentrated neighborhood unemployment. This is essentially Bartik's prescription for urban unemployment.

A second approach involves affecting the intrametropolitan demand for labor. That is, maintaining or increasing labor demand in and around higher unemployment neighborhoods. It is this set of policies and programs that are of concern here. In particular, this book aims to discern whether increasing the number of jobs near higher unemployment neighborhoods is likely to improve the employment status of neighborhood residents.

Besides metropolitan and neighborhood-targeted approaches, there are other submetropolitan strategies, including those that target central cities as a whole. But these two approaches represent the extreme ends of labor demand approaches to the problem. There differences extend beyond geographic scale and frequently involve distinct political contexts, with metropolitan policies viewed as trickle-down or *corporate center* approaches and neighborhood-targeted policies viewed as equity-based or *alternative* approaches (Robinson, 1989). Corporate center approaches are focused on increasing overall investment and jobs in the region, a rising tide that will lift all boats. Alternative approaches often stress job creation at the neighborhood level with special focus on high-unemployment areas. The *Chicago Works Together* plan of the Harold Washington Administration in Chicago is among the best known examples of alternative, decentralized economic development planning that focused on neighborhood-targeted economic development (Mier and Moe, 1991).

Goetz distinguishes development agencies who rely on Type I economic development policy, in which policy is aimed wholly at increasing the level of economic activity in a locality or region, from those employing Type II policy, in which policies are aimed at maximizing benefits for residents of targeted areas or for certain income groups (Goetz, 1990). Type I policy is essentially concerned with overall growth, while Type II policy is more concerned with the distribution of growth and development, both across space and among individuals. While the literature varies on how economic development policy can be distinguished, there is a common distinction between supporting aggregate economic growth throughout a central city or metropolitan area, on the one hand, and targeting efforts to increase jobs and economic activity in and for the benefit of lower-income areas on the other. Neighborhood-targeted economic development policies generally fall into the Type II approach to economic development. They seek to reduce unemployment in higher unemployment neighborhoods by spurring jobs in and near these areas. They may also seek to bring other benefits to targeted neighborhoods.

At the federal level, HUD is the principal federal agency that implements neighborhood-targeted economic development policies. HUD administers CDBG, much of which is targeted to residents of lower-income, urban neighborhoods. CDBG is used to fund business development activity, including business financing, technical assistance, and neighborhood planning functions (Unites States Department of Housing and Urban Development, 1995a). The fiscal year 1995 budget for CDBG was $4.8 billion. HUD also administers the new federal Empowerment Zone program, which provides labor- and capital-based incentives for firms located in six urban subcity areas nationally for business expansion or location, as well as for hiring of zone residents. Firms located in Empowerment Zones are eligible for tax credits of up to $3,000 per zone resident employed. The program also includes access to accelerated depreciation and tax-exempt bond financing for private projects. Beyond incentives to firms, the Empowerment Zone program provides up to $100 million to each zone that can be used over ten years for a wide variety of potential uses in the zone, including job training and business development. The program also allows zone administrative bodies to request waivers of federal regulations for activity occurring within the zone (United States Department of Housing and Urban Development, 1995b).

Other federal policies that seek to increase the number of jobs in and around distressed urban neighborhoods include those in the Commerce Department, the Treasury Department, and elsewhere. In the Commerce Department, the Economic Development Administration provides capital for local revolving loan funds that target areas with high levels of unemployment, including rural areas and urban neighborhoods. In late October, 1995, the Treasury Department issued a request for proposals for the Community Development Financial Institutions Fund, a $50 million dollar program to capitalize and support efforts aimed at financing community development projects in distressed rural and urban areas, including those with high unemployment and poverty rates. Eligible activities include the financing of projects aimed at nearby job creation for residents of distressed neighborhoods (United States Department of the Treasury, 1995).

State and local efforts to increase jobs in and around urban neighborhoods vary widely across the United States. Enterprise zones are among the most popular approaches. Erickson reports on a survey of 37 states administering 39 enterprise zone programs that provided some combination of supply-side subsidies, typically for capital, land or labor, to firms in distressed areas, many of them urban neighborhoods (Erickson, 1992). Job creation was mentioned as a goal by more than half of the programs, and was often linked either explicitly or implicitly to zone residents. Zone incentives included reductions in taxes on plant and equipment, below market-rate financing, and tax credits for job creation.

Enterprise zone programs are not the only way that states can target economic development assistance to neighborhood areas. Some states support business and community development financing programs that target subcity areas. An example is Connecticut's Urbank program, which provides portfolio insurance to banks making loans to small businesses in targeted areas, with increasing levels of insurance for more distressed areas. The State matches borrower contributions to the participating bank's special reserve fund that can be used to cover loan losses under the program (Connecticut Development Authority, 1994). Other states have made grants to nongovernmental programs whose purpose includes the economic development of distressed areas. The State of North Carolina, for example, has made grants totaling $2 million to the Center for Community Self-Help in North Carolina to capitalize a business development financing program that targets rural areas and distressed urban neighborhoods.

Municipal governments employ a wide variety of approaches to neighborhood-targeted economic development, including business financing programs, industrial park development, tax increment financing, and other approaches. Robinson classifies 139 municipal economic development and planning departments as adopting one of three approaches to targeting economic development assistance to geographic areas or underserved communities: *supportive, guiding,* or *supportive and guiding* (Robinson, 1989). *Supportive* implies practice in which the agency does not attempt to encourage development in one area over another, while *guiding* implies a strong role in steering development assistance to targeted areas. While only 17 of the 139 agencies classified themselves as purely *guiding*, another 91 classified themselves as *supportive and guiding*. Thus, 78 percent of the agencies described themselves as directing at least some of their efforts toward targeted communities.

While some neighborhood-targeted economic development programs provide subsidy only after residents of targeted areas are hired, others only require firms receiving subsidy to use "best efforts" to hire residents of targeted areas. Still others do not attempt to intervene in the hiring process at all. Of the 39 enterprise zone programs examined by Erickson, for example, at least 12 did not have a job tax credit that was tied to hiring local residents or low-income individuals, and only 11 tied other incentives for capital or land to local hiring or hiring low-income individuals (Erickson, 1992).

There remains a great deal of policy activity in the arena of submetropolitan- and neighborhood-targeted economic development programs that seek to improve the employment status of high unemployment neighborhood residents. The continuing notion that intrametropolitan space can be a barrier to employment buttresses such policies. Of course, improving housing access among lower-income residents to new job centers and increasing the transportation mobility of these residents are other frequently prescribed solutions. Regardless of the plusses and minuses of these solutions, there remains some question whether job access at the neighborhood level is significant in affecting neighborhood unemployment and what other factors are significant or interacting. By examining the effects of different neighborhood characteristics, including nearby job opportunities, on unemployment and nearby working, one can better understand the potential for neighborhood-targeted economic development efforts to achieve their employment status objectives.

NOTES

1. There are 77 community areas in the city of Chicago (the central city). Community areas are aggregations of census tracts that range in total 1990 population from 3,314 to 114,079, with most having populations of 20,000 to 60,000. The 45 low- and moderate-income areas are those that had median family incomes at or below 80 percent of the median income for the Chicago metropolitan area in 1990 ($41,745).

2. Jobs data are from *Where Workers Work*, Illinois Department of Employment Security, as compiled in Immergluck, 1994. Unemployment data are from compilations of decennial U.S. Census data, STF3A, contained in Immergluck, 1994.

3. Included in west side zip codes are 60608, 60612, 60622 - 60624, 60635, 60639, 60644, 60647, and 60651.

CHAPTER 2

Theory and Evidence on the Impacts of Neighborhood Economic Development Policies

THEORETICAL CRITIQUES OF PLACE-BASED ECONOMIC DEVELOPMENT

To address the issue of urban unemployment, a variety of strategic policy orientations might be adopted. Ladd suggests three basic policy approaches that can be identified for dealing with distressed urban neighborhoods (Ladd, 1994). First, people-oriented strategies focus on helping people but do not give attention to helping the places where they live. Second, place-based people strategies focus on a neighborhood or a larger geographic area to increase the economic well being of people living in the area. Third, pure place-based strategies aim to improve the physical and economic vitality of a geographical place without explicit attention to the people who live in the place.

Economic development policies are place-based—either pure place-based or place-based people, in Ladd's typology—policies that are intended to alter private investment and employment decisions in specific regions, states, metropolitan areas, cities or neighborhoods. The objectives of these policies include increasing the employment and earnings prospects of residents in the target area, diversifying and increasing the tax base, and providing local commercial services and amenities. While place-based economic development strategies have been employed in the United States, they have been more popular in Europe. For example, Great Britain had designated "Special Areas"—targeted development areas with high unemployment rates—dating back at least to the late 1930s (Hill, 1977). The enterprise zone concept also originated in Great Britain, often attributed to Sir Geoffrey Howe, who became Margaret Thatcher's Foreign Secretary (Butler, 1991).

The argument of place versus people is an old one. Typically, economists have been among the chief advocates for people-based development strategies. Urban planners, sociologists and human geographers have more often been advocates for giving attention to urban places as well as people. They often view neighborhood space, in particular, as representing community, especially for lower-income populations.

The earliest discussion in the literature of the place versus people debate is often attributed to Louis Winnick, who titled the debate, "place prosperity versus people prosperity" (Winnick, 1966). The conflict is between improving the welfare of people as individuals, regardless of where they live, and improving the welfare of groups or communities defined by geographical proximity. Winnick argues that policies that attract economic activity to certain places do so at the expense of other areas. This is essentially the zero-sum argument that Bartik addresses (Bartik, 1991). Winnick adds that place-based policies are inefficient and ineffective, with much of the benefits going to other than the intended beneficiaries.

> . . . the incremental income occasioned by planned intervention is subject to considerable leakage and flows as readily into the pockets of the least needful as into the pockets of the most needful. (Winnick, 1966)

One reason place-based policies are inefficient, from Winnick's perspective, is that aid becomes conditional on persons continuing to live in a declining place. Another problem is that prosperous areas contain distressed households, just as distressed areas contain some prosperous households.

Kain and Persky characterize place-based economic development policies focused on distressed urban areas as inefficient methods for improving the welfare of disadvantaged individuals and households (Kain and Persky, 1969). Indeed, the bulk of the economics literature is concerned with the welfare of the individual or household alone, and not of the community.

Logan and Molotch offer a radical perspective on local place-based economic development policy (Logan and Molotch, 1987). They describe the city as a growth machine, politically dominated by coalitions of large, locally dependent firms that are the driving forces of urban economic development in the United States These coalitions are organized to preserve the concentration and accumulation of wealth among the corporate elite.

State and local economic development efforts are fundamentally inequitable and ineffective, according to Logan and Molotch. First, they suggest that local economic growth does not make jobs, but only redistributes them. They, like Winnick, see local and regional economic development as zero sum, with exogenous product demand determining overall employment levels. Logan and Molotch argue that near-perfect labor mobility will ensure that as jobs are created in one location, workers from other locations will follow to obtain those jobs. They also claim that increases in local jobs are likely to go to outsiders rather than the indigenous workers in whose name economic development is pursued.

Logan and Molotch also argue that local economic growth will benefit owners of fixed factors of production, land and embedded capital. Owners of such capital are dependent on local economic growth to reap higher monopoly rents. The benefits of local economic growth are seen as flowing, as Winnick writes, "into the pockets of the least needful."

THEORETICAL ARGUMENTS FOR PLACE-BASED POLICIES

Advocates of place-based policies respond to criticisms of economic development in a number of ways. Some suggest that local or regional economic growth will benefit disadvantaged workers. Slack labor markets are seen as harming those at lower ends of the income spectrum because they are the "last hired and first fired." Bartik uses this logic in suggesting that disadvantaged workers and the unemployed may have the most to gain from local and regional economic growth (Bartik, 1991). Bartik adds that such benefits may accrue through internal labor markets, i.e., through promotional advancement at existing employers. He also suggests that, even if growth provides only a short-run shock to labor demand, temporary employment of previously unemployed individuals will have a permanent hysterisis effect. By this, he means that the work experience provided from the demand shock will provide permanent human capital benefits to workers with limited work experience. In a separate article, Bartik concedes that many new jobs go to inmigrants from other areas, but argues that local growth still benefits local residents (Bartik, 1993).

Bartik counters Winnick and Logan and Molotch on their critique of local economic development as a zero-sum endeavor. He suggests that a redistribution of labor demand may be efficient if it increases job opportunities in areas that have higher unemployment, where the marginal utility of employment is greater (Bartik, 1991).

In considering enterprise zone programs, Papke argues that targeted economic development may have positive employment benefits for local residents (Papke, 1993). She suggests that, if zones are small relative to the rest of the economy, the effect on zone wages and employment will depend on the elasticity of supply of factors to the zone and on the elasticity of demand for zone output. Subsidies to labor, or to labor and capital uniformly, should increase wages paid by zone firms. Employment effects will depend on the elasticity of labor supply in the area.

Zone subsidies should generally increase zone production, creating an output-effect increase in the demand for labor. Subsidies to capital alone, however, may result in a substitution of capital for labor, and local employment will increase only if the output effect exceeds the substitution effect. Then, assuming employment does increase, jobs may be filled either by residents within or outside the targeted area. Bartik's work at the regional level, however, suggests that 75 percent or more of new jobs resulting from regional growth may go to inmigrants (Bartik, 1993). At the neighborhood level, an even higher percentage of new jobs would be expected to go to nonzone residents, most of whom are not faced with the barrier of intermetropolitan relocation. So the employment benefits from zone development might not benefit zone residents very much.

Papke notes that the labor supply of disadvantaged or unskilled workers may be more elastically supplied than for workers in general. Thus, nearby job growth may induce more workers into the workforce. She finds that a subsidy to all labor employed in the zone always increases employment and wages at zone establishments. She also finds that a capital subsidy reduces zone employment and wages at low elasticities of product demand and low labor supply elasticities. If product demand is completely inelastic, an equal increase in subsidy to capital and labor will reduce zone employment and wages. When product demand is elastic, and only resident wages are subsidized, equal subsidies to capital and labor significantly affect employment and wages.

Bolton provides another set of arguments for place-based economic development policies (Bolton, 1992). He argues that such policies may help preserve a sense of community in places where it already exists. Place and community remain important social values that provide a specific context for personal interaction.

Bolton explains that, to defend place-based policies, at least one of three different arguments must be adopted. First, one can challenge the conclusion that place-based policies cause great inefficiencies. He suggests second-best arguments that question whether these policies are introduced into a perfectly competitive economy. Second, one can maintain that place-based policies are actually necessary to improve efficiency, because place-specific market imperfections make such intervention desirable. This is similar to Bartik's defense of place-based economic development, in which he suggests that markets do not recognize the higher marginal utility of employment in areas with relatively higher unemployment (Bartik, 1991). Third, one can challenge the notion that place-based policies are not efficient at redistributing income to individuals.

Bolton uses welfare economics to develop some of these arguments further. He argues that rigid wages in declining areas may overstate the true opportunity costs of utilizing local human capital. He also suggests that governments and utilities that use average cost pricing will overprice capital in declining areas because marginal cost is likely to be below average cost. In the case of both human and infrastructure capital, market prices may exceed true opportunity costs, leading to market inefficiency.

Bolton's main point is that a place can possess a sense of community that is also a form of place-based capital. This community capital is productive, and residents who have a strong sense of place know and appreciate it. It is an intangible, location-specific asset. Bolton's concept of community capital is reminiscent of Jane Jacobs' notion of neighborhood vitality (Jacobs, 1961). Without large amounts of activity, neighborhoods are doomed to become "perpetual slums." She argues that neighborhoods with activity will eventually work their way out of economic hardship, unless they are blocked from doing so through redlining, housing discrimination, or the actions of developers and city planners. It is neighborhood vitality, a sort of community capital, that enables a neighborhood to "unslum."

The importance of a sense of place in urban neighborhoods suggests that people-oriented policies are incomplete because they ignore the value of Bolton's community capital and Jacob's neighborhood vitality. By separating individuals and households from the communities in which they live, such policies assume that place means nothing to the individual or to the community. These policies fail to recognize the value of each individual to his or her community. If policies assist people in leaving their communities, or encourage them to do so, they may impose an unseen cost on both the individual and his or her community. Wilson argues this point in suggesting that desegregation policies imposed costs on black communities by unleashing a pent-up demand by black middle-class households for suburban and nonblack neighborhoods (Wilson, 1987).

Similarly, policies aimed at assisting residents of low-income neighborhoods find employment in distant parts of the metropolitan region, without any concern for the quality of life in these neighborhoods, may exacerbate the spatial and social isolation of those lower-income households who remain in the neighborhood with few employment opportunities. This is not to argue that such programs are inherently misguided, but that they may require complementary efforts to maintain or improve the quality of life in the distressed neighborhoods. Hughes adopts such a line of thought in arguing for complementing job mobility policies with increased crime prevention programs in lower-income neighborhoods (Hughes, 1989).

As suggested in Chapter 1, neighborhood economic development and the location of economic activity may have impacts beyond employment effects. Wood argues, for example, that the emergence of abandoned, dilapidated industrial sites provides evidence of the disastrous implications of urban industrial decline to residential community life in affected neighborhoods (Wood, 1991). These buildings, by discouraging nearby investment and providing locations for illicit activities, can destroy a neighborhood's sense of place, i.e., its community capital.

EVIDENCE ON THE IMPACT OF NEIGHBORHOOD ECONOMIC DEVELOPMENT POLICY ON FIRM ACTIVITY, JOB GROWTH AND THE EMPLOYMENT OF NEIGHBORHOOD RESIDENTS

The majority of work on the effects of neighborhood economic growth has focused on its effect on land values and on overall jobs, without attention to the nature of the jobs and who obtains them. This is due in part to the fact that evaluations have been focused on the impact on places and firms, but not on the people or communities associated with these places. It is also partly due to a lack of data on employees of firms in targeted areas.

Most studies on intrametropolitan effects of public policy on economic growth have focused on the impacts of taxes and fiscal factors on the location, investment and employment levels of firms across central cities and suburban areas. Unlike the mixed findings of intermetropolitan studies, there appears to be strong evidence that fiscal factors do affect intrametropolitan growth. Substantial evidence exists to suggest that property taxes do affect the site decisions of firms, for example. Bartik cites seven studies since 1979, among which the average long run elasticity of a jurisdiction's industrial activity with respect to its property tax rate is -1.76 (Bartik, 1992). Thus, a 10 percent decrease in the property tax rate, from 2 to 1.8 percent for example, would result in a 17.6 percent increase in business activity. Bartik claims, however, that few of the jobs will be obtained by the original residents of the jurisdiction, at least in small areas, but presents no evidence for this latter assertion. Less attention has focused on the effect of public services on intrametropolitan growth. Charney finds that the proportion of land served by water and sewer is a highly significant determinant of the number of firms locating in an area with a 0.48 elasticity (Charney, 1975). She finds highway proximity to be significant only at the 10 percent level.

Evaluations of state enterprise zone programs provide evidence on the effects of policies that are explicitly intended to spur development in distressed areas. These studies have tended to focus on impacts on firm behavior and job growth, and only secondarily on whether residents of zones attain employment. In an evaluation of seven enterprise zones in seven different cities, Dabney compares growth rates in the number of establishments in the zones before designation to growth rates after designation (Dabney, 1991). He finds that growth rates increased substantially for small, independent businesses, but that this increase was not significantly different from the change in growth in the rest of the corresponding cities.

Erickson analyzes data from a HUD survey done in 1985 and 1987, where zone coordinators were asked to report on program performance (Erickson, 1992). The survey covered 357 zones in 186 communities. Median performance indicators, as reported by zone administrators, were as follows:

Jobs created: 175
Investment: $4.5 million
New establishments: 3
Percent of jobs to low- or moderate-income individuals: 44

Erickson does not provide comprehensive data on program costs, only anecdotal reports suggesting that benefits exceeded costs. Erickson finds that zones with large numbers of incentives do affect zone business activity and that the programs are less effective in states with large numbers of zones.

Studies of the cost-effectiveness of enterprise zones range from a General Accounting Office study of Maryland's program, which finds that the program was not successful at creating any jobs and credited all contemporaneous job gains to other factors, to a study of the Evansville, Indiana enterprise zone, which finds a cost effectiveness of $1,633 per job (United States G.A.O., 1988; Rubin and Wilder, 1989). The Evansville study finds that the cost per zone-resident-job was $43,579, and a study of all Indiana enterprise zones finds cost per zone-resident-job to be $53,506 (Indiana Department of Commerce, 1992). None of these studies attempt to measure the duration of employment of residents hired by zone establishments.

In evaluating the Indiana enterprise zone program, Papke finds that zone designation appears to have a positive impact on the local labor market (Papke, 1993). Unemployment claims at local unemployment offices declined by 19 percent. Areas served by unemployment offices, however, did not match zone boundaries, so that effects on zone residents could not be precisely measured. By examining census data, Papke ascertains that residents in zone areas did not see significant increases in economic well-being after zone designation. She does not employ quasi-experimental control tracts or areas, however, thereby providing no isolation of zone designation effects.

CHAPTER 3

Theory and Evidence on Factors Affecting Unemployment and Employment Patterns across Urban Space

THEORIES ON BARRIERS TO EMPLOYMENT AND JOB-WORKER MATCHING ACROSS URBAN SPACE

Economic Theory

There are two general bodies of economic theory that inform examinations of urban unemployment at the neighborhood level. The first is the area of labor economics, including theories of human capital and skill mismatch, segmented labor markets, and discrimination. These phenomena are important to problems of urban unemployment, but much of the general theory is not well suited to intraurban labor market analysis.

The second area of relevant economic theory is urban economics, particularly recent work addressing urban unemployment issues. Rigorous theoretical work here is scarce, while empirical work is quite plentiful. Although scarce, such theory is useful for identifying characteristics of labor supply and demand that affect the relationship between urban economic growth and the employment of neighborhood residents.

One body of research particularly relevant to urban unemployment is the spatial mismatch literature. Kain is generally credited with introducing this topic (Kain, 1968). His three hypotheses are: 1) residential segregation affects the distribution of black employment; 2) residential segregation reduces black employment opportunities; and 3) postwar suburbanization has seriously aggravated the problem. After describing the high levels of residential segregation of blacks from 1940 to 1960, Kain suggests several reasons why residential segregation affects the distribution and level of black employment. These include:

a)   the distance to and difficulty of reaching certain jobs from black neighborhoods may impose costs on blacks large enough to discourage them from seeking these jobs;

b)   blacks may have less information on, and less opportunity to learn about, jobs distant from their homes or those of their friends; and

c)   employers located outside the ghetto may discriminate against blacks out of real or imagined fears of retaliation from white customers for bringing blacks into all white residential areas. Conversely, businesses in the ghetto may discriminate in favor of blacks.

As Ihlanfeldt explains, a), above, leads to blacks being more likely to work in or near their neighborhoods than whites, essentially Kain's first hypothesis (Ihlanfeldt, 1992). Interestingly, much of the economics literature that responds to Kain's original proposition appears to downplay the role of information and networks, as well as the role of direct or indirect workplace discrimination. This is due partially to the fact that Kain's empirical work does not address which of these mechanisms are important or to what degree.

While Kain clearly sees residential segregation as the key driver of the problem of black employment access, the mechanisms b) and c), above, may not be resolved by desegregation alone, at least not unless desegregation reaches a critical mass. The relocation of a single black household into a white suburban neighborhood may actually exacerbate information problems, by taking away the little information that the household did have in its former community and not necessarily offering a viable replacement. Also, workplace discrimination may continue and black applicants may find nearby jobs more prevalent but less accessible.

It is Kain's third hypothesis, the aggravation of the problem by the decentralization of employment and the inability of blacks to follow the jobs, that has formed the center of the spatial mismatch literature. Ihlanfeldt suggests that a more general statement of Kain's third hypothesis is that the spatial mismatch between where blacks live and where jobs are located reduces the net annual earnings of central-city blacks.

Holzer expands Kain's explanation of how segregation hurts black employment prospects (Holzer, 1991). Search costs will be higher and information networks may be less effective for those spatially isolated from jobs. He adds that these disadvantages will be more important for those living farther from these jobs.

The decline of jobs in central-city areas is key to the spatial mismatch hypothesis. Johnson and Oliver point out that the spatial distribution of employment is affected by two general phenomena: deindustrialization and deconcentration (Johnson and Oliver, 1992). The former is the shift in employment in regional economies from manufacturing to other sectors. Deconcentration is the intrametropolitan migration of establishments out of central-city neighborhoods. Both can contribute to spatial mismatch, while deindustrialization alone contributes to skills mismatch.

Aspatial labor market theory of relevance to urban unemployment includes that of skills mismatch, discrimination, and segmented labor markets. Bluestone and Harrison describe the shift in the United States economy from highly unionized, high wage manufacturing to high technology manufacturing, services, and other industries, where wages are not as high (Bluestone and Harrison, 1982). Wilson argues that urban blacks have been particularly vulnerable to these structural economic changes, especially the shift from the goods-producing sector, with its relatively low skill requirements, to the service-producing sector (Wilson, 1987). Kasarda argues that both skill and spatial mismatches have been key factors in reduced employment opportunities for central-city blacks (Kasarda, 1993).

Conventional neoclassical assumptions about labor markets distinguish between pure and statistical discrimination. In the former, employers have a taste for discrimination and will pay a premium to avoid members of a discriminated group. Statistical discrimination occurs when employers use race or ethnicity, or some other trait, as a signal of productivity aspects that are costly to measure. Neoclassical theorists argue that, in the long run, both pure and statistical discrimination will be eliminated under perfect competition (Becker, 1957).

The segmented labor market hypothesis rejects the market equilibrium assumption of neoclassical labor economics (Cain, 1976). Workers in career track jobs are protected because their jobs are not subject to competition from outside the firm. Jobs in this segment are rationed, not auctioned. Thus, unlike the conventional neoclassical framework, discrimination and unequal access to employment can persist.

In focusing particularly on unemployment in cities, Harrison argues against the neoclassical notion that workers are paid according to the marginal productivity of their human capital (Harrison, 1972). He argues that productivity, itself, is an increasing function of wages, so that higher pay induces greater productivity. He also suggests that many workers are relegated to low-productivity and low-pay jobs regardless of their ability by a system of dual labor markets. Harrison distinguishes between primary and secondary labor markets. The primary market, or core sector, is dominated by oligopolistic employers who are able to provide training and capital to support jobs offering high wages. The secondary or peripheral market is composed of highly competitive, labor-intensive firms that do not invest in training and pay low wages. These firms are able to thrive without investing in job-specific training and do not depend on a stable work force. Due to historical discrimination and the exclusion of minorities from the networks of those employed in the core sector, minorities are left to fill the peripheral sector jobs, while whites have access to core sector employment. It is important to note that this research is somewhat dated, preceding large scale industrial restructuring, growth of affirmative action programs, and changes in public sector employment during the 1970s and 1980s.

The Island Economy Framework for Understanding Urban Unemployment

A comprehensive theoretical treatment of the relationship between urban spatial structure and urban unemployment is provided by Simpson, who develops a theory of spatial mobility to explain urban unemployment problems (Simpson, 1992). He argues that worker mobility in urban areas is not costless and that these costs are important to understanding urban labor markets.

Simpson asserts that the metropolitan areas consist of a series of urban subdivisions or local labor market islands. Mobility among the islands is costly in both direct (travel costs) and indirect (job information) terms. He assumes that all workers prefer the local labor market of their place of residence. More skilled workers are assumed to adopt more formal and spatially extensive job search strategies. This is due to the sparseness of suitable job opportunities across islands. While direct job search opportunity costs for high-skilled workers are high due to high earnings, returns to the search are also large. The acceptance of a job offer depends upon the attractiveness of a job, represented by the wage offer.

According to Simpson, when local labor market demand is high, local wage offers are high as firms compete for relatively scarce local workers. Workers accept these offers and periods of unemployment are short. Thus, unemployment will be lower when local labor demand is strong. The effect of local labor demand on unemployment depends upon the skill level of the job opportunities. Wage offers for higher-skilled jobs are less sensitive to local labor demand than are offers for lower-skilled jobs.

Simpson begins with a model of individual unemployment probability for a certain time interval as:

$$p = p(G,s) \text{ with } dp/ds < 0, \ d(dp/ds)/dG > 0 \tag{1}$$

where G represents a series of dummy variables representing different skill levels, G_i, $i=1 \ldots k$, and s is the number of jobs per resident on the island. Let Z be a vector of resident characteristics such as sex and age, and $g(Z_i)$ be the probability of a local labor force participant satisfying characteristic Z_i, $i = 1 \ldots n$. If $g(Z_i)$ and the probability of employment, p, are independent, then the unemployment rate on any island may be expressed as:

$$U(G, s; Z) = U_0(s) + \sum_{i=1}^{n} U_i g(Z_i) + \sum_{i=2}^{n} U_{n-1+i} G_{i-1} \tag{2}$$

where U_0 represents the rate of unemployment owing solely to local labor market conditions and the U_i are components that represent the increased unemployment arising because individuals satisfy characteristics Z_i or fall into skill categories G_i.

This specification permits a linear regression model of unemployment rates across islands to be formulated as:

$$U = d_0 + d_1 s + d_2 s^2 + \sum_{i=1}^{n} d_{i+2} Z_i + \sum_{i=1}^{n} d_{n+2+i} G_i + e \qquad (3)$$

$U_0(s)$ is expressed as a quadratic because, besides choosing nearby locations, workers from more distant locations may search for jobs in locations with high values of s as these locations provide many job opportunities. This implies that, as s gets very large, it overestimates the probability of receiving an acceptable job offer in the area, so:

$$dU/ds < 0 \qquad\qquad \text{but} \qquad\qquad dU/ds^2 > 0 \qquad (4)$$

This implies that $d_1 < 0$ and $d_2 > 0$.

In Simpson's island economy model, workers require information not only about general labor market conditions for the city and beyond, but also about the distribution of conditions across intraurban space. This may result in intraregional conflicting trends. In periods of rapid economic expansion, local firms are more likely to expand their operations and, therefore, to relocate, particularly out of central cities, where facilities are often outdated. Consequently, generally good economic times may adversely affect labor demand in many central-city labor islands.

Sociological and Geographic Theories

Wilson argues that highly concentrated poverty in some central-city neighborhoods is both a cause and effect of urban unemployment (Wilson, 1987). While he suggests that structural economic changes have been key to increasing joblessness among urban black males, he also develops a rich framework of neighborhood breakdown in which neighborhood residents withdraw from the labor force in large numbers. This joblessness is then compounded by extreme social and spatial isolation from the remaining segments of the metropolitan community, including working- and middle-class blacks who have become more residentially mobile.

Social isolation not only deprives inner-city neighborhood residents of resources and conventional role models, it also denies them the social learning derived from mainstream networks that facilitate social and economic advancement in modern industrial society. So Wilson sees the lack of social mobility, caused in part by a lack of spatial mobility among poor blacks and an increased spatial mobility among the middle-class blacks, as the key problem. With this social isolation comes adverse social behavior including lower marriage rates and increased criminal behavior.

Massey and Denton argue that residential hypersegregation of blacks was the driving force of black joblessness during the 1970s and 1980s (Massey and Denton, 1993). They agree with Wilson that structural economic change played a role in creating the urban underclass, but argue that residential hypersegregation made it a black underclass. The changes in urban economies harmed many racial and ethnic groups, they argue, but only black residents were highly segregated so that the resulting income loss was confined to a small set of contiguous and racially homogeneous neighborhoods. The largest drops in consumer demand arising from layoffs, lower wages, and increasing unemployment were confined to neighborhoods inhabited exclusively by poor blacks. This explains the decline in retail and personal service establishments, although it does not directly explain the loss in manufacturing and other types of businesses.

Hanson and Pratt maintain that neighborhood space constitutes an important parameter in the decisions of both employers and workers in the job search and hiring process (Hanson and Pratt, 1992, 1995). Employers prefer workers from nearby neighborhoods and, therefore, utilize neighborhood newspapers and word-of-mouth to identify job candidates. The residential rootedness of many workers means that they seek out local employers. In combination with employers' preferences for nearby workers, this local preference creates distinctive, "microscopic" labor markets.

Granovetter offers two general criticisms of the neoclassical approach to labor market analysis (Granovetter, 1992). He argues that neoclassical approaches often rely on an exaggerated version of methodological individualism, in which individuals and households are atomized from the influence of their relations with others, from others' decisions and behaviors, and from the history of these relations. He suggests that, in addition to resulting in an incorrect understanding of how labor markets function, such approaches make it difficult to describe how individual actions aggregate up to the level of institutions. The mechanisms of social networks and relationships are typically ignored. Institutions are assumed to arise due to rational functionality alone and to suit the objectives already being described and estimated. Granovetter argues that the economic literature generally fails to consider intertwining economic and noneconomic motives, including sociability, approval, status, and power. These motivations may prove more powerful than traditional economic motives.

Granovetter describes how a worker develops relationships with coworkers who become aware of the worker's abilities and personality as he or she moves through a series of jobs. As workers develop longer lists of such acquaintants, their access to job opportunities improves. The power of such networks grows as the number of jobs held by workers or the number of jobs held by workers' past and present coworkers increase. According to Granovetter, higher-skilled workers rely more on these weak-tied networks than lower-skilled workers, who rely more on strong-tied networks consisting of family and friends. Thus, residential location may be important to the effectiveness of networks at providing employment access. This is consistent with Wilson's concentrated poverty theory.

O'Regan and Quigley, like Simpson, recognize that mobility costs include both transportation and information costs (O'Regan and Quigley, 1991). They suggest that information costs may be the greater barrier to employment and that members of Wilson's isolated underclass may simply not know very many people who can be of much help in finding work. This social isolation deprives people from access to information networks that would lead to employment opportunities. Members of networks that include many employed people have superior employment prospects.

Discrimination is another area where sociologists offer theory relevant to urban unemployment. Kirschenman and Neckerman argue that, with few exceptions, most considerations of labor market racial discrimination have neglected the processes at the level of the firm that underlie discrimination effects (Kirschenman and Neckerman, 1991). They note that a taste for discrimination in employment practices may lead to perceived and actual productivity differences between groups, making statistical discrimination more likely. Expectations about group differences in productivity may bias job placement decisions and evaluations of job performance. Such expectations can result in discriminated groups receiving less on-the-job training. Kirschenman and Neckerman also argue that productivity is a social phenomenon, not an individual characteristic, depending on the social relations of coworkers, customers, and others. If a firm's dominant customer has a taste for discrimination, for example, the firm may adopt a similar preference to keep productivity high.

EVIDENCE ON BARRIERS TO EMPLOYMENT

The empirical literature on barriers to employment that is most relevant to neighborhood unemployment and the tendency of neighborhood labor forces to attain nearby jobs addresses at least one of the following categories of factors: 1) the spatial mismatch between jobs and residents; 2) skill-based factors, including mismatches between new jobs and resident skills; 3) employment discrimination; and 4) networks. Much of the literature addresses more than one of these mechanisms as compounding factors, although the focus is typically on only one or two. I make no attempt to review the entire literature on factors affecting unemployment; I focus on the research most relevant to urban neighborhood unemployment.

Spatial Barriers

Ihlanfeldt reviews 30 articles that address intraurban spatial barriers to employment, often referred to as the spatial mismatch literature (Ihlanfeldt, 1992). In addition to Ihlanfeldt's, there have been at least three relatively comprehensive reviews of this literature (Jencks and Mayer, 1990; Holzer, 1991; and Kain,1992).

Jencks and Mayer survey more than 25 studies on the spatial mismatch subject. They conclude that, together, these studies provide no clear evidence supporting Kain's overall hypothesis. Holzer concludes that spatial mismatch is a significant contributor to black-white employment differences but that the magnitude of the contribution is unclear. Kain reviews both Jencks and Mayer's and Holzer's reviews, discusses in detail those original studies that appear to have been most influential, including those of Ellwood and Leonard, and concludes that spatial mismatch is significant (Ellwood, 1986; Leonard, 1986).

Ihlanfeldt's survey is the most systematic in that it categorizes the bulk of the literature into six methodological approaches: microdata comparisons of central-city and suburban residents; regressions of black economic welfare on job decentralization or housing segregation; comparisons of wage rates paid by work location; the use of a direct measure of accessibility; interracial comparisons of commuting times and distances; and the use of establishment-level data. He then adds a seventh category of literature focused on youth employment.

Ihlanfeldt finds consistent support for Kain's first hypothesis, that residential segregation affects the geographical distribution of black employment. He argues that there is much less consensus on both Kain's second hypothesis, that housing segregation reduces black economic welfare, and his third hypothesis, that there exists a mismatch between where blacks live and where jobs are located. After Ihlanfeldt dismisses studies that he maintains are flawed methodologically, however, he finds strong and consistent support for Kain's second and third hypotheses.

Ihlanfeldt's own work is among the most advanced methodologically (Ihlanfeldt, 1992). Working with microdata from Public Use Microsamples, he looks primarily at the effect of commute time on both black and Hispanic youth. He attempts to address factors not covered by much of the economics literature. For example, he tries to address the network and concentration effects suggested by Wilson (Wilson, 1987). Wilson argues that social isolation of very poor youth may impair their employment prospects. Ihlanfeldt attempts to control for this by separately regressing the effect of distance on employment for youth living in poverty. He finds that the effect of distance is significant for this population, and is nearly identical to the general effect for all youth.

Overall, Ihlanfeldt finds that the effect of a reduction in travel time of 5 minutes increases the probability of employment by 7 percentage points for white youth, and 5.5 percentage points for black and Hispanic youth. Calculated at the mean employment rate for each group, this same reduction will cause a 17 percent increase in the employment rate for whites, a 19 percent increase for Hispanics, and a 29 percent increase for blacks. Ihlanfeldt attributes 27 percent of the difference in employment rates between white and black youth to inferior physical access, and an even higher 35 percent of the difference between Hispanic and white youth.

Note, however, that Ihlanfeldt's use of the microdata does not explicitly address Wilson's theory. Wilson argues that social isolation is caused by the spatial concentration of poverty, not by the poverty of households alone. Appropriately addressing Wilson's theory in Ihlanfeldt's methodology would require incorporating some measurement of the level of poverty of the neighborhood of the individual, which is not provided in the microsample data.

Since these comprehensive reviews were written, additional research has been published on spatial mismatch questions. While some consensus appears to have developed that spatial barriers to employment can be significant, the debate continues, especially over the magnitude of the effects and the appropriate specification of job access measures. In their analysis of 1990 Detroit data, Bauder and Perle (1995) find that spatial job access has only a minimal effect on job access, and Carlson and Theodore (1997) find very small effects of job proximity on earnings in examining 1990 PUMS and zip code level jobs data for Chicago. In examining data for the San Francisco area, Raphael (1995) finds substantial employment rate effects due to changes in the number of jobs near the residential location of workers. Rogers (1997) uses zip code job access measures for the Pittsburgh area to find, like Raphael, that an increase in the number of jobs near workers' places of residence reduces unemployment spells. Rogers considers only job levels or job growth, and not resident population densities, in her specifications of job access, thus not accounting for how many other residents are competing for nearby jobs. None of these recent studies seeks to measure the impact of jobs near a neighborhood at a radius consistent with neighborhood economic development policy. Moreover, while some consider general occupational levels and industrial mix, none explicitly controls for the occupational mix of nearby jobs vis-à-vis residents.

The relative importance of changes in job levels versus the raw levels of jobs remains ambiguous. While it may be that the changes in the number of jobs affect the number of employment opportunities in an area, large variances in job densities across urban areas should result in neighborhoods with more nearby jobs also having many more gross job openings at local firms over a given time period. Holzer (1996) finds that gross openings and gross hires are driven by employee turnover, not net employment growth at firms. Of course, net growth affects gross hires, but because annual turnover runs on the order of 20-

25 percent of total employment at a firm (Holzer, 1996), even substantial growth or loss of employment might not be as important as the raw level of jobs in determining gross hires. Annual net job growth or loss is likely to be modest compared to annual turnover rates. For the decade from 1979 to 1989 on Chicago's industrial predominantly black and West Side, which bore a disproportionate share of the city's job loss, the decline in jobs amounted to only 22 percent over the decade, or less than an average of 2.5 percent per year.

Gross hires are likely to be an important measure of nearby labor demand because as the number of gross openings increases, the opportunity for matches to arise between nearby jobs and neighborhood residents should increase. Even if an area loses a substantial number of jobs, if many jobs remain in the area, regular employee turnover will still create many openings, providing a relatively high level of appropriate job matches with nearby residents. This is not to say that net employment growth should have no effect. Rather, it seems that both raw level and net growth should be expected to have some effect, and that raw level may be more important given the magnitude of annual employee turnover at firms.

Skill Barriers

Kasarda considers the spatial deconcentration of jobs across metropolitan areas to be a significant contributor to increasing unemployment among lower-skilled urban residents, particularly blacks (Kasarda, 1993). However, he points to the structural change in metropolitan economies, from goods-producing to information and service industries, as another key cause of increasing urban unemployment during the 1970s. Over the 1970s and 1980s, manufacturing employment in most large northeastern and Midwestern metropolitan areas declined as a proportion of all jobs, and declined in absolute terms in central cities. At the same time, service jobs, including higher-skilled information-based jobs increased, especially in newer suburban areas. Thus, the overall shift in national and regional employment from manufacturing to services—often termed deindustrialization—combined with job flight to the suburbs to form a doubly hard blow to low-skilled central-city residents, particularly blacks. Low education levels among urban blacks combined with decreasing numbers of central-city jobs requiring a high school education or less to result in poorer employment prospects for urban blacks. Looking at nine northeastern and Midwestern central cities, Kasarda finds considerable growth from 1950 to 1970 in the number of employed blacks who had not completed 12 years of schooling. From 1970 to 1980, however, these cities saw a drop in poorly educated black males who worked full-time and an increase in the percentage not working. For Chicago, Kasarda's calculations show that the number of job holders in the central city with less than a high school degree declined by 41 percent over the 1970s. By 1980, only 23 percent of central-city jobs in Chicago were held by those with less than a high school diploma. At the same time, 45 percent of black male residents and 58 percent of unemployed black male residents did not have a high school diploma.

Wilson stresses a growing skill mismatch as the key structural factor in increasing urban unemployment (Wilson, 1987). He suggests that blacks' weaker connections to the labor force put them in the position of accepting the brunt of the adverse impacts of structural changes. Blacks were last hired and first fired as the economy changed, and their lower educational achievement left them ill-prepared to adapt quickly to acquire new job skills.

Holzer (1996) uses a survey of employers in four large cities to determine the hiring needs of firms and the problems of skill deficiencies among job applicants. He finds that employers are increasingly requiring math and reading skills, as well as computer skills, for entry level jobs. He also finds that the lack of such skills is a major problem for many less-educated, and especially, minority job applicants.

Employment Discrimination Barriers

Ellwood is well known for suggesting that labor market discrimination is the key barrier to employment for minority residents of central-city neighborhoods, particularly for black youth (Ellwood, 1986). There is widespread consensus that racial discrimination plays some significant role in access to employment, especially for urban blacks. Even those who argue that spatial and skill barriers are important, frequently concede that race, including discrimination, is a very important factor in urban unemployment. Kasarda suggests:

there is no question that race, including outright discrimination, plays
a potent—and probably the most powerful—role in the relatively
poor performance of blacks . . . (Kasarda, 1993)

According to data from the 1990 General Social Survey, 65 percent of
whites characterized blacks as lazier than whites, and 56 percent rated
blacks less intelligent than whites (Peterson and Vroman, 1992).

Bates uses the 1987 Census Characteristics of Business Owners
survey to find that white-owned small businesses employ primarily
white workers (Bates, 1993). Even white-owned firms in minority
areas employ largely white workforces. Meanwhile, black-owned firms
employ largely minority workers, regardless of location. Bates finds
that 89 percent of black-owned firms in the survey have workforces
that are at least 75 percent minority, while only 18 percent of white-
owned firms had workforces that were at least 75 percent minority. Of
white-owned firms in minority areas, only 29 percent employ more
than 75 percent minority workforces. Networks as well as
discrimination may account for these differences.

To measure racial discrimination directly, experimental audits
must be used. An example of experimental evidence is presented by
Turner et al. in a matched-pair audit of several hundred employers in
two metropolitan areas (Turner et al., 1991). They find that black job
applicants with comparable employment histories and skills to those of
white applicants received fewer employment offers than the white
applicants. Whites received offers for 29 percent of applications while
blacks received offers for only 19 percent.

Another approach to detecting likely discrimination is from
interview and survey research. Kirschenman and Neckerman
interviewed 185 employers in the Chicago area on the issue of hiring
and race (Kirschenman and Neckerman, 1991). They found that
employers did not hesitate to generalize about racial or ethnic
differences in workforce quality. They also found that employers used
concepts of race, space and class interchangeably, linking "black" with
"inner-city" and "lower-class" in many cases. However, in describing
their own practices, it was common for employers to describe a more
heterogeneous workforce that included both good and bad workers.

Kirschenman and Neckerman also found that employers preferred
Hispanic workers, particularly Mexican-American workers, to blacks.
Hypothesizing that discrimination would be evidenced by black-

Hispanic hiring differences for jobs where no basic (e.g., language) skills were required, they compared the percentage of those jobs filled by blacks and Hispanics. They found that, for city employers, almost 43 percent of no-skill jobs went to Hispanics and only 22 percent went to blacks, even though low-skilled blacks generally suffer from higher unemployment rates than do low-skilled Hispanics. For jobs requiring basic skills, however, the trend was reversed, with 39 percent of such jobs going to blacks and 19.5 percent to Hispanics. Interestingly, these trends were reversed for suburban employers, with employers hiring many more blacks for no-skill jobs and many more Hispanics for basic skill jobs.

Kirschenman and Neckerman suggest that, other things being equal, employers prefer hiring Hispanic workers over blacks. In the central city, employers find Hispanics in relatively short supply for jobs requiring basic skills, but in the suburbs this is not the case. This suggests that the supply of Hispanic workers might be appropriately viewed as highly segmented, with a group of better-skilled, desirable workers employed more in the suburbs and a segment of immigrant laborers lacking basic language skills working more in the central city.

Using firm survey data, Holzer (1996) finds that firms hire black job applicants at substantially lower rates than either white or Hispanic applicants. Because Hispanic applicants have substantially lower educational attainment levels, discrimination is suggested. Holzer also finds that suburban and smaller firms hire blacks at lower rates than central-city and larger firms.

Network Barriers

A variety of research has examined the effect of formal and informal networks on employment. Based on the theories of Wilson and Granovetter, various types of social networks might be important factors in access to employment (Wilson, 1987; Granovetter, 1992). Granovetter's work suggests that about half of all jobs are obtained through informal networks. Granovetter's weak-tied networks may be less relevant to urban unemployment than the strong-tied networks of family and friends that are of more concern to Wilson, because weak-tied networks are more critical to the job searches of higher-skilled workers, who rely on their contacts with many occupational acquaintances. In the case of the lower-skilled residents of central-city neighborhoods, friends, relatives and neighbors, rather than previous occupational acquaintances, may be more important to employment prospects.

O'Regan and Quigley examine the 1980 Public Use Microsample data on more than 55,000 youths in 47 large metropolitan areas (O'Regan and Quigley, 1993). They utilize the employment status of parents and siblings as proxies for access to informal, strong-tied networks. They find that, after controlling for metropolitan-wide influences, youths are far more likely to be employed if parents or siblings are employed. They are also likely to have the same industrial affiliations as their parents and are likely to be employed in the same part of the metropolitan area. These findings are important in indicating the importance of strong-tied networks but do not directly address the entirety of neighborhood effects about which Wilson is concerned.

In a study of workers and employers in the Worcester, Massachusetts area, Hanson and Pratt find that the vast majority (94.3 percent) of employers they interviewed preferred to hire through word-of-mouth (Hanson and Pratt, 1992). The employers suggested that this method enabled them to hire people who "fit in" the existing company culture. Hanson and Pratt also find that those who find work via word-of-mouth or informal means work closer to home than those who find work through more formal channels. For women, those finding work informally traveled an average of 15.3 minutes to work, while those finding work through formal channels traveled an average of 19.4 minutes. The results are also significant for men, with informal job finders traveling 14.8 minutes and formal job finders traveling 18.3 miles. Of course, formal job seekers may tend to be higher-skilled workers; this is consistent with the notion that higher-skilled workers conduct geographically wider job searches.

CHAPTER 4

Developing Models of Neighborhood Unemployment and Local Working Rates

THE EFFECT OF JOBS ON NEIGHBORHOOD UNEMPLOYMENT RATES AND ON LOCAL WORKING

In this chapter, I develop a model to estimate the effect of near-the-neighborhood labor demand on two neighborhood employment status conditions: 1) neighborhood unemployment; and 2) the proportion of neighborhood residents in the labor force employed in or near the neighborhood. I call the latter the "local working rate." Most recent examinations of the determinants of unemployment utilize data that have little detailed geographic information. If labor market demand is included as an independent variable, it is often captured only at the metropolitan level. One body of research that attempts to incorporate the effects of submetropolitan labor market demand on individual- or neighborhood-level employment status is the spatial mismatch literature reviewed in Chapter 3. Kain, Ellwood, and Leonard, in particular, employ neighborhood level employment status and job access measures to estimate the effect of job access on unemployment (Kain, 1968; Ellwood, 1986; Leonard, 1986). Key shortcomings of this work, however, include imprecise geographic constructs of neighborhood and nearby job markets as well as the failure to account for occupational mixes of residents or jobs. More recently, spatial mismatch work has focused on the use of metropolitan microsample data, but these data do not provide neighborhood-level information on jobs and their proximity to residents.

No previous research has looked at the effect of nearby labor demand on the proportion of neighborhood residents employed near the neighborhood. Some work has been done on developing models to estimate the residential originations of new workers when employment is added at a particular job site. Soot and Sen construct a gravity model that predicts the residential origin and racial and ethnic makeup of new workers when jobs are created during specific economic development projects (Soot and Sen, 1991). This model, however, does not comprehensively address which factors affect the degree to which residents of different types of neighborhoods work near their neighborhood.

The models developed here resemble most closely the approach of Simpson, reviewed in Chapter 3 (Simpson, 1992). Simpson examines the effect of submetropolitan labor demand in London and Toronto on submetropolitan unemployment rates. There are three key features of Simpson's approach that make it insufficient for addressing the questions at hand here.

First, Simpson's geographical unit is too large to add to the understanding of the *neighborhood effects* of concern here. His unit of analysis has an average population of approximately 250,000, comparable to substantial portions of many large central cities. The relatively small (less than 50) number of these large areas in his data sets leaves him with relatively few degrees of freedom for regression purposes.

The second problem with Simpson's approach is that he uses the same geographic unit to compare jobs and the employment status of residents. Unemployment rates vary widely over small neighborhood areas, due in large part to differences in labor supply characteristics. In cities that are highly segregated over small areas, choosing neighborhood units that are too large obfuscates labor supply differences. At the same time, neighborhood residents seeking and finding nearby jobs are more likely to find work in a larger, surrounding area than within the small neighborhood unit. Thus, a better geographic scheme for examining employment status utilizes very small, relatively homogeneous residential areas together with larger surrounding job catchment areas. A key advantage of this approach is that it reduces any concern with the causality of jobs vis-à-vis unemployment. That is, local or neighborhood unemployment rates may affect the employment and location decisions of firms across metropolitan space. By utilizing a residential unit much smaller than the surrounding job area, the characteristics of each particular neighborhood are unlikely to affect, in a substantial way, the total job levels in the larger surrounding area, thus minimizing the potential endogeneity problem.

Finally, Simpson incorporates relatively little characteristic data on residents and workers into his research. The lack of information on the racial makeup of residents is perhaps the most problematic.

MODELING NEIGHBORHOOD UNEMPLOYMENT

Given that search and commute costs across metropolitan space may be significant, especially for those in lower-skilled occupations, youth, and those seeking part-time employment, the presence of nearby jobs may affect not only the propensity to attain work locally, but also the unemployment rate. I first develop a model for neighborhood unemployment and then apply the same basic model to local working.

The potential for nearby labor demand to affect neighborhood unemployment should be especially strong in the case of large, densely populated metropolitan areas with significant traffic congestion such as Chicago, where commute times can be substantial. In the nine-county Chicago area, approximately 11 percent of workers commuted at least one hour each way to their jobs in 1990, another 11 percent commuted at least 45-59 minutes, and 23 percent commuted from 30 to 44 minutes.[1] Job search costs may also pose significant barriers to employment across metropolitan space, especially in racially and economically segregated urban areas. In these areas, the knowledge of jobs in distant locations may be scarce, especially for lower-skilled workers. Finally, racial discrimination by employers may be confounded with spatial segregation, with employers in nonminority areas preferring nonminority employees, either based on their own discriminatory preferences or on those of their customers. The latter may be particularly relevant in retail and consumer service industries.

In simplest form, neighborhood unemployment might be expected to decline as the number of jobs within a nearby job catchment area rises. A spatial model that assumes a circular job catchment area

extending a constant radius from the neighborhood center is shown in
Figure 1. A simple functional form for such a model is:

$$u = \alpha + \beta J_{(d)} \qquad\qquad (5)$$

where

u = unemployment rate of the neighborhood;

$J_{(d)}$ = number of jobs within a certain radius, d;

and

$\beta < 0$ is expected.

This is a fairly simple geometric scheme for the job catchment
area. More sophisticated approaches might incorporate a radius that is a
function of neighborhood characteristics such as access to mass transit
or expressways or some measure of traffic congestion. Other
approaches might assume a noncircular catchment area. Such
extensions of the model will not be considered here.

If proximity to jobs improves employment prospects for
neighborhood residents as this model implies, nearby jobs will be
accessible to residents of the nearby job catchment area as well. Thus,
more persons living in the nearby job area increases the competition
neighborhood residents face for these jobs. For example, if 1,000 jobs
and 500 residents in the labor force lie within two miles of
neighborhood A, while 1,000 jobs and 20,000 residents lie within two
miles of neighborhood B, residents of neighborhood A would be
expected to have greater access to nearby jobs than those of
neighborhood B. Much of the competing labor force in the surrounding
area is likely to live closer to many of the nearby jobs than do
neighborhood residents. This suggests that the key measure of a
neighborhood's accessibility to nearby jobs is not the number of jobs
alone, but the ratio of jobs to the number of persons in the labor force
within radius d—the nearby jobs-to-labor-force ratio. A neighborhood
with a nearby jobs-to-labor-force ratio of one-half, for example, would

**Figure 1. The Basic Geographical Scheme for a Nearby Job
Catchment Area Surrounding a Neighborhood.**

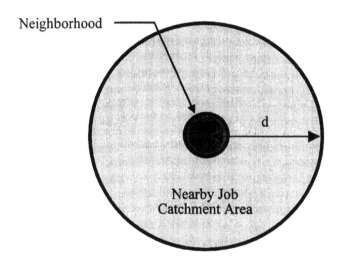

be expected to offer fewer nearby job opportunities to neighborhood residents than a neighborhood with a nearby jobs-to-labor-force ratio of two.

So a better model of the effect of nearby job accessibility on unemployment is:

$$u = \alpha + \beta(J_{(d)}/R_{(d)}) \tag{6}$$

where

$J_{(d)}$ = number of jobs within a radius, d, of the neighborhood;

$R_{(d)}$ = the number of residents in the labor force within a radius, d, of the neighborhood; and

$J_{(d)}/R_{(d)}$ = nearby jobs-to-labor-force ratio.

Based on the literature on job search, workers in higher-skilled occupations conduct more spatially expansive job searches than workers in lower occupational levels (Simpson, 1992; Granovetter, 1992). Therefore, nearby labor demand is not expected to affect the employment prospects of workers in high-skilled occupation, and a modification of (6) leads to:

$$u = \alpha + \beta(J_{LM(d)}/R_{LM(d)}) \tag{7}$$

where

$J_{LM(d)}$ = the number of jobs in low- and moderate-skilled occupations within distance, d;

$R_{LM(d)}$ = the number of residents in low- and moderate-skilled occupations in the labor-force (including the unemployed) within distance d; and

$J_{LM(d)}/R_{LM(d)}$ = nearby jobs-to-labor-force ratio.

A more complete set of unemployment determinants is drawn from the literature. Key factors should include measures of the average occupational skill level and race of residents in the labor force, and the similarity between occupations of the nearby jobs and the occupations of the neighborhood labor force. The average occupational level of nearby jobs may also prove important, based again on the notion of a wider geographic job-resident match process for higher-skill jobs. Thus, the general model with which to estimate neighborhood unemployment rates is:

$$u = \alpha + \beta(J_{LM(d)}/R_{LM(d)}) + \gamma(x_1, \ldots x_k) \qquad (8)$$

where $x_1, \ldots x_k$ is a vector of characteristics of neighborhood residents, nearby jobs, and occupational similarity between the two.

CHOOSING THE NEIGHBORHOOD AREA AND NEARBY JOB RADIUS

The model presented in equation (8) and illustrated in Figure 1 requires two types of geographic units. First, the smaller residential neighborhood unit must be chosen. Because the model incorporates both labor supply and labor demand factors, and because labor supply characteristics, including race and skills, are likely to vary greatly over relatively small distances, small neighborhood units are most appropriate. Census tracts, or even smaller census block groups, are typically used when analyzing neighborhood phenomena such as segregation, housing conditions, and poverty distributions. To ensure adequate variances for independent variables and a large number of observations for model estimation, residential areas should be chosen as small as is practical. The lower limit is determined primarily by maintaining sufficient population in each observation to permit the characterization of neighborhoods across a wide variety of features (e.g., employment status, industry, occupation, enrollment, etc.) and by the minimum disaggregation level of the available data. In the data set used here, the smallest unit is a half-mile-by-half-mile area commonly called a quartersection, which I adopt as the neighborhood unit.

While unemployment rates can vary over very small neighborhood areas, neighborhood residents seeking and finding nearby jobs are more likely to find work in some larger, surrounding area. The model illustrated in Figure 1 presents a circular job catchment area of radius d surrounding and including the neighborhood. Jobs within this catchment area are considered nearby. Adopting a job catchment area that is substantially larger than the neighborhood itself mitigates problems of jobs-unemployment causality. The job catchment area can be treated as exogenous to the much smaller neighborhood unit at its core.

It is also important to choose a job catchment radius that is consistent with neighborhood economic development policy and practice. Most neighborhood economic development efforts in the United States serve areas on the order of five to 25 square miles. Assuming a job catchment area that is circular, this suggests a radius of between 1.3 and 2.8 miles. The City of Chicago, for example, has identified 22 manufacturing areas throughout Chicago's central city, which have been officially designated as industrial districts. Some of these stretch for several linear miles and cover several square miles. The City targets economic development programs to these areas. It also uses federal Community Development Block Grant and job training funds to support the activities of neighborhood-based economic and labor force development groups to provide technical assistance to local firms and help firms find local residents for jobs. These organizations also raise funds from foundations, state government, and banks and corporations. Typically, service areas of these efforts range from five to 20 square miles.

Given the focus on neighborhood-targeted economic development efforts, I adopt a two-mile radius around each residential quartersection. The spatial job search literature reviewed by Simpson suggests that a radius of two miles is not an inappropriate size when considering low- and moderate-income workers (Simpson, 1992). He points to research that workers are more likely to travel more than 15 miles and less likely to travel less than 2.5 miles the higher their skill levels.

The approach adopted here captures the labor supply differences among urban neighborhoods, provides a large degree of freedom in statistical calculations, recognizes the radial nature of surrounding labor markets, and obviates concerns about job-unemployment causality. It suggests that residents seeking and attaining nearby employment will most often find it in areas surrounding, but not directly within, their own small neighborhood. This geographic approach also provides a consistent unit for both neighborhood and nearby job catchment areas rather than comparing areas with varying geographic shapes or sizes.

THE MODEL FOR LOCAL WORKING RATES

In addition to the goal of reduced unemployment, a secondary objective of neighborhood economic development policy is to provide nearby jobs for the labor force of the neighborhood. Residents who do not have to commute great distances to find work are better off. For those who are employed in low-wage or part-time work, long commute times might make it hard to maintain employment in the long run. Besides shorter commutes, increasing the number of residents employed near the neighborhood may have substantial long-term benefits for the neighborhood, including access to strong job networks and good relations between residents and local firms. Residents may not always be supportive or accommodating of local employers due to negative externalities such as traffic and pollution. If a large portion of residents work at nearby employers, however, they are likely to be more supportive of neighborhood economic development efforts.

Many of the factors that are expected to affect neighborhood unemployment might be expected to affect the proportion of neighborhood residents in the labor force employed within a certain radius of the neighborhood, or the "local working rate." Certainly, the nearby jobs-to-labor-force ratio would be expected to be positively related to residents working nearby. The real question is to what degree. Also, some labor supply and demand characteristics may be significant determinants of local working or unemployment but not both. For example, the gender make-up of the labor force might be expected to influence local working more than unemployment if one suspects that women prefer to work near home to a greater degree than men. So the general model for estimating the local working rate is:

$$w = \alpha + \beta(J_{LM(d)}/R_{LM(d)}) + \gamma(x_1, \ldots x_k) \tag{9}$$

where

> w = the number of neighborhood residents employed within a radius divided by the number of neighborhood residents in the labor force;

and

> $x_1, \ldots x_k$ are the same variables as in equation (8).

NOTE

1. Based on 1990 Census Transportation Planning Package, CTPP-1, U.S. Census Bureau. The CTPP Urban Element for Chicago includes Cook, DuPage, Lake, McHenry, Kane, Kendall, Grundy, Will and Kankakee counties, all in Illinois.

CHAPTER 5

Neighborhood-Level Job and Demographic Patterns in Chicago

THE DATA

The data used to estimate the models of equations (8) and (9) in Chapter 4 come from the 1990 Census Transportation and Planning Package (CTPP) Urban Element for Chicago. The CTPP, which is derived from the 1990 Census long form survey of the population, provides three sets of data (referred to as CTPP-1, CTPP-2 and CTPP-3) on residents and workers aggregated at small geographic levels. A data dictionary for the CTPP is found in the Appendix.

CTPP-1, sometimes referred to as the place-of-residence file, provides aggregate information on the residents of quartersections, half-mile-by-half-mile areas, within the nine-county Chicago metropolitan area. Data in this file include many standard census fields including population, age, sex, ethnicity by race, occupation, industry, enrollment in school, median income, median earnings, etc. While some cross tabulations are provided (e.g., sex by age), these are limited. Because human capital derived from formal education may be an important characteristic, it would be desirable to have a variable that provided some measure of this feature. Unfortunately, the CTPP provides only the enrollment status of residents by age group (12 to 17 years and 18 to 64 years), and not educational attainment data. It is likely, however, that the enrollment status of 12 to 17 year-olds is highly correlated to the educational quality and attainment of young adults. Enrollment rates for 12 to 17 year-olds correspond directly to drop-out rates and so are likely to be correlated to the quality of local schools. School quality, in turn, may be more important to predicting employment status of young adults than years of schooling, given the diversity of educational quality across metropolitan space.

CTPP-2, sometimes referred to as the place-of-work file, provides aggregate information on the workers employed in the metropolitan area by the quartersection of workplace. Fields include race, ethnicity, occupation, industry and other characteristics of workers.

CTPP-3, often called the origination-destination file, provides information on the commutes from each quartersection to every other quartersection to which any resident commutes for work. These data include mode of transportation and commuting time of day. Characteristics such as race, ethnicity, occupation, and industry are not provided for commuting workers in CTPP-3.

AGGREGATING THE DATA AND CONSTRUCTING THE PRINCIPAL DATA SET

The basic geographical structure used to estimate the models in equations (8) and (9) of Chapter 4 follows the scheme of Figure 1, adapted to correspond to the CTPP quartersection scheme for Chicago and to a defined nearby area radius of two miles. The nearby job catchment area contains 49 quartersection units totaling 12.25 square miles.[1] In the study area for this research, these catchment areas contain an average of 17,880 jobs with a median of 11,206. They have a mean population of 64,814 with a median of 42,946.[2] The quartersection, essentially a half-mile-by-half-mile area, is the unit at which the Chicago CTPP data are disaggregated and so are used as the residential neighborhood unit. These areas are roughly similar in population to census tracts, with a mean of 1,500 and a median of 1,020 for the study area. Many of these quartersections have very small populations, however, with 25 percent having a population of less than 360 persons.

From the CTPP, a data set was constructed for a large, central portion of the Chicago metropolitan area, including the entire central city and most of surrounding Cook and DuPage Counties as well as a portion of northwestern Will County. The study area contains a wide variety of neighborhoods, including low- and high-income areas, areas near low and high job densities, and suburban and central-city areas.

For every residential quartersection in the study area, variables describing jobs and residents within two miles of the quartersection (including the jobs and residents in the quartersection itself) were calculated. The total study area consists of almost 3,300 residential quartersections. All of the quartersections in the study area are not included in the analyses, however; several sets of exclusions were performed prior to forming the principal data set. A large number of these quartersections have relatively small populations, with more than 1,500 cases having fewer than 500 persons over 15 years old in the labor force. Because the meaningful calculation of many neighborhood residential characteristics of concern (e.g., unemployment rate, proportion of working-age persons who are under 25 years old, proportion of residents who are black, occupational level index, etc.) requires a substantial population and labor force, cases with fewer than 500 persons in the labor force are excluded from the analysis. This leaves approximately 1,700 residential quartersections in the study area.

Another three types of quartersections were excluded. Sixty-nine quartersections that lie within two miles of the Chicago central business district (CBD) are removed from the analysis.[3] Some of these quartersections have very high local working rates, i.e., large portions of the resident labor force work within two miles of the neighborhood, because the CBD is within this range. The effect of CBD jobs on neighborhood unemployment rates are not the concern here, however, so these observations are excluded. Second, sixteen quartersections that are estimated to contain more than 200 units of public housing are excluded from the analysis.[4] These quartersections exhibit extreme unemployment rates (many over 30 percent, with some over 50 percent) and/or extraordinarily large portions of the adult population not in the labor force. Most of these neighborhoods turn up as outliers in analyses of the data set, including estimations of models explaining unemployment rates. Finally, three remaining cases where there are no youths aged 12-17 are eliminated so that a variable describing the proportion of teenagers enrolled in school can be included in the analysis.

The resulting principal data set consists of 1,645 cases and includes variables describing residential characteristics for each of these quartersection neighborhoods, as well as variables describing jobs and residents in the labor force within two miles of the neighborhood. Residents of these quartersections who are in the labor force constitute 80 percent of the labor force in the study area and 53 percent of the labor force in the nine-county Chicago metropolitan area. These quartersections also represent approximately 81 percent of the low- and moderate-skilled labor force residing in the study area and 56 percent of low- and moderate-skilled labor force in the nine-county area.[5]

For each independent, dependent or descriptive variable, Table II provides a description of the variable and summary statistics for the principal data set. Table II shows that the principal data set includes a large number of low-, middle- and upper-income neighborhoods.

SUMMARY STATISTICS AND THE SPATIAL DISTRIBUTION OF DEPENDENT AND KEY INDEPENDENT VARIABLES

Neighborhood Unemployment Rates

Table II shows that, for the principal data set, the March, 1990 neighborhood unemployment rate varies from 0 to 43 percent, with a mean of 7 percent, a median of 5 percent and a standard deviation of 6 percent. Twenty-five percent of neighborhoods in the principal data set have unemployment rates of less than 3 percent, and 25 percent have rates above 8 percent. Approximately 3 percent of the cases have unemployment rates of 0 percent. Figure 2 illustrates the spatial distribution of unemployment rates for residential neighborhoods in the principal data set.[6] Neighborhoods with very high unemployment rates, exceeding the mean of 7 percent plus two standard deviations (19 percent total), are clustered on the west and south sides of the central city. Similarly, neighborhoods with unemployment rates between one (13 percent) and two (19 percent) standard deviations above the mean are generally clustered in these same parts of the metropolitan area.

Neighborhoods farther from the central city's CBD tend to fall into lower unemployment rate categories. One clear exception to this is the cluster of very high unemployment neighborhoods in predominantly black parts of the southern Cook County suburbs (coordinates: $Y = 10$; $X = 24.5$).

Table II. Summary Statistics for Dependent, Independent and Descriptive Variables (1,645 Observations)

Variable Description	Variable Name	Mean	Median	SD*	Min.	Max.
Unemployment Rate	Unemployment	0.07	0.05	0.06	0.00	0.43
Proportion of Labor Force Working Within 2 Miles	Local Working Rate	0.15	0.15	0.08	0.00	0.60
Low- and Moderate-Skilled Jobs Within 2 Miles / Low-and Moderate-Skilled Labor Force Within 2 Miles	Jobs-to-Labor-Force	0.87	0.75	0.52	0.02	5.54
Occupational Level Index for Employed Residents	Occupational Level	10.80	10.80	0.89	8.12	13.42
Occupational Dissimilarity Between Employed Residents and Nearby Jobs	Occupational Dissimilarity	0.19	0.19	0.06	0.05	0.51
Proportion of Residents Who are Black	Proportion Black	0.18	0.01	0.33	0.00	1.00
Proportion of Residents Who are Hispanic	Proportion Hispanic	0.08	0.03	0.15	0.00	1.00
Proportion of Residents Age 16-64 Under 25	Working-Age Residents Under 25	0.19	0.18	0.06	0.03	0.68
Proportion of Residents in the Labor Force Who are Female	Female Labor Force	0.46	0.46	0.05	0.29	0.63

Table II. Continued

Variable Description	Variable Name	Mean	Median	SD	Min.	Max.
Proportion of Residents Age 12-17 Enrolled in School	Teenage Enrollment	0.94	0.96	0.07	0.45	1.00
Proportion of Residents Age 18-64 Enrolled in School	Adult Enrollment	0.13	0.12	0.05	0.02	0.66
Occupational Level Index for Jobs Within 2 Miles	Job Occupational Level	10.94	10.94	0.26	8.47	12.36
Distance from Chicago CBD in Miles	Miles From CBD	15.55	14.04	7.27	3.00	36.46
Population		2,378	1,726	1,732	724	14,177
Median Household Income($)		39,186	37,574	14,381	6,514	113,687
Low- and Moderate-Skilled Jobs Within 2 Miles		17,847	14,995	10,874	57	61,700

*SD = Standard Deviation

**Figure 2. March 1990 Unemployment Rate for
Residential Quartersections in the Principal Data Set.**

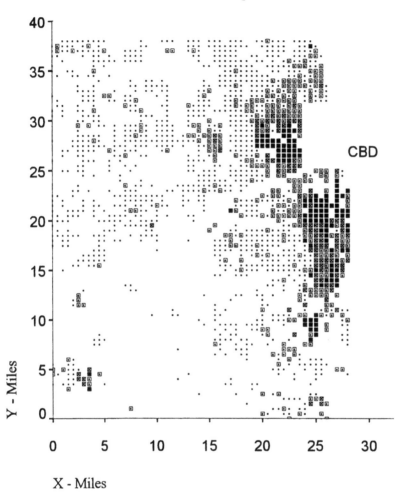

Unemployment Rate

■   >0.19

▨   0.13 - <=0.19

▣   0.07 - <=0.13

·   <=0.07

The Local Working Rate

Besides understanding the geographic patterns of unemployment rates, those interested in neighborhood economic development should understand what affects the tendency for neighborhood residents to attain work near the neighborhood. The local working rate, w, is the proportion of residents in the labor force employed within two miles of the neighborhood. Local working rates vary from 0 to 60 percent in the principal data set, with a mean of 15 percent, a median of 15 percent and a standard deviation of 8 percent. Twenty-five percent of observations in the data set have local working rates of less than 10 percent, and 25 percent have values of more than 20 percent. Approximately 2 percent of the neighborhoods in the principal data set have local working rates of 0 percent.

Figure 3 illustrates the varying levels of local working rates for the quartersections in the principal data set. Areas with large local working rates include: the western Cook County and DuPage County corridor lying south and southwest of O'Hare Airport and north of Interstate 290 ($X = 2$ to 18; $Y = 28$ to 31); parts of the southwest Cook County suburbs ($X = 18$ to 21; $Y = 15$ to 19); and Joliet ($X = 2$; $Y = 4$). Neighborhoods with very low local working rates include many on the southern and western parts of the central city.

Local working rates can be compared to the proportion of *employed* residents (versus labor force) working within two miles, or *w*. There is a very high correlation (0.9929) between w and *w*. In some cases where unemployment is high, w is substantially lower than *w*, but most quartersections with low w also have low *w*.

Jobs-to-Labor-Force Ratio

The first independent variable of interest is the nearby jobs-to-labor-force ratio, J_{LM}/R_{LM} in equations (8) and (9). This is equal to the number of low- and moderate-skilled jobs within two miles of the neighborhood divided by the number of persons in the labor force who live within two miles of the neighborhood and are not employed in high-skilled jobs.[7] This is the principal measure of spatial access to nearby jobs. This variable alone, however, does not measure the occupational level of the jobs in the surrounding area, nor does it

Figure 3. March 1990 Local Working Rate: Proportion of Residents in the Labor Force Working Within Two Miles of a Quartersection.

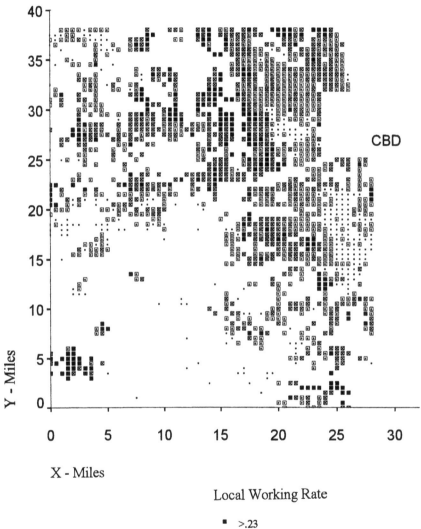

Local Working Rate

■   >.23

▨   >.15 - <=.23

▣   >.07 - <=.15

·   <=.07

measure the differences between the occupations of nearby jobs and the occupations of neighborhood residents. These characteristics will be addressed through separate independent variables discussed below.

In the principal data set, the jobs-to-labor-force ratio varies from 0.02 to 5.54 with a mean of 0.87, a median of 0.75, and a standard deviation of 0.52. Twenty-five percent of the 1,645 observations have jobs-to-labor-force ratios of less than 0.56, and 25 percent of the observations have ratios of greater than 1.05. Figure 4 plots the jobs-to-labor-force ratio for observations in the study.

While jobs have been leaving the central city over the last thirty years, especially on the south and west sides, 1990 job densities surrounding many central-city neighborhoods are still much higher than those surrounding most suburban neighborhoods. (This is not the case for much of the central city's south side.) However, many of these neighborhoods are also surrounded by large low- and moderate-skilled labor forces, thus decreasing their jobs-to-labor-force ratios. Even among the neighborhoods on the west side of the central city that have suffered substantial population losses, most still contain and/or are located near large labor forces. Neighborhoods in the low-income, predominantly black North Lawndale community area (X=21 to 22; Y =26 to 27), for example, have lost substantial population but are within two miles of highly populated Hispanic neighborhoods such as the Little Village and Pilsen community areas to the south and southeast, where the labor force is stable or increasing. Thus, the nearby jobs-to-labor-force ratios for such neighborhoods are not very large. Nearby labor force levels are so high in many central-city quartersections that the jobs-to-labor-force ratios are relatively low. This effect is exacerbated in many neighborhoods on the south side where job levels are low but nearby population densities are still relatively high.

Conversely, job densities in many suburban areas are not as high as in many central-city neighborhoods, but relatively low population densities combined with the relatively high skill levels of local residents in these newer areas result in much higher jobs-to-labor-force ratios than in the central city.

Figure 4. March 1990 Jobs-to-Labor-Force Ratio (J_{LM}/R_{LM})

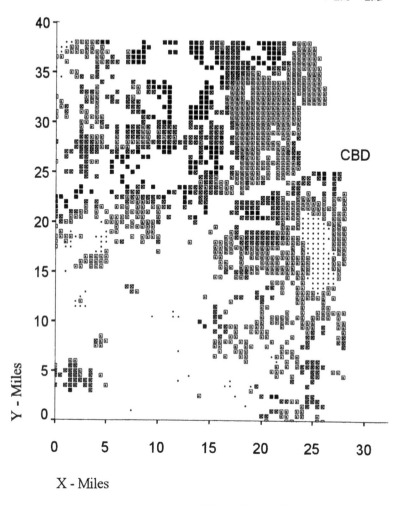

Jobs-to-Labor-Force

- **■** >1.39
- ▨ >0.87 - <=1.39
- ▫ >0.35 - <=0.87
- · <=0.35

Figure 4 can be compared to Figure 2 to identify any obvious relationships between the jobs-to-labor-force ratio and neighborhood unemployment. A visual comparison suggests a negative but fairly weak relationship between the two variables; this is corroborated by a correlation coefficient of -0.27. Low and very low values of nearby jobs-to-labor-force on the central city's south side generally correspond to quite high unemployment rates in many of these neighborhoods, and many areas with high jobs-to-labor-force ratios have below average unemployment. There are, however, many clear exceptions to this pattern. One example is a cluster of western Cook County suburbs (e.g., the Maywood area, $X=16$; $Y=28$) that exhibit relatively high unemployment rates (between 7 and 19 percent) and yet have large nearby jobs-to-labor-force ratios.

Figure 4 can also be compared to Figure 3 to discern any obvious patterns between the jobs-to-labor-force and local working variables . The relationship here is positive, as indicated by a correlation coefficient of 0.43. Areas near O'Hare Airport and in the western suburbs with high jobs-to-labor-force ratios also tend to exhibit high local working rates. At the same time, south side neighborhoods in the central city with very low jobs-to-labor-force ratios exhibit very low local working rates. West side central-city neighborhoods ($X=20$ to 22; $Y=27$) whose jobs-to-labor-force ratios are not very low also suffer from low local working rates. However, west side local working rates are lower than those of many neighborhoods with similar jobs-to-labor-force ratios on the north and northwest sides of the city ($X>20$; $Y=30$ to 35), suggesting other factors may be at work. The west side neighborhoods are predominantly black, so race may be playing a role here.

Occupational Level Index

A variable expected to be important to both neighborhood unemployment and local working is the occupational level of employed neighborhood residents. A scalar index was created to measure the average occupational level of each neighborhood's employed residents. Median hourly earnings for each occupation were used as scalar weights. Median hourly earnings, including wage and salary plus self-employment, were calculated for all workers in Cook and DuPage counties—roughly equivalent to the study area—from the Public Use

Microsample of the 1990 Census. Median wages should provide a good estimate of the skill level of each occupational grouping. An index with a potential range from 15.38 (hourly earnings in dollars for professionals) to 4.17 (hourly earnings in dollars for armed forces) was calculated as follows:

Occupational Level Index =

{15.38(number employed in professional specialty occupations)
+15(number of employed residents in executive, administrative, and managerial occupations)
+13.46(number of residents employed in precision production, craft and repair occupations)
+12.02(number in protective service)
+11.42(number in transportation, moving, etc.)
+11.24(number of employed residents in technician and related support occupations)
+9.80(number in sales)
+9.05(number in administrative support, including clerical)
+8.65(number of machine operators/assemblers)
+6.73(number in farming, forestry and fishing)
+6.53(number in service occupations except households)
+4.90(number of handlers, equipment cleaners and laborers)
+4.37(number in private household)
+4.17(number in armed forces)}/ Number of employed residents (10)

For example, a neighborhood in which all employed residents are executives would have an occupational level index of 15, while a neighborhood where one-half of the employed residents were machine operators and one-half were precision production workers would have an index of $1/2(8.65)+1/2(13.46) = 11.06$. The actual range of the occupational level index for the principal data set is from 8.12 to 13.42. The mean value of 10.80 corresponds to an occupational level between that of the technician and sales categories. The median value is 10.80 and the standard deviation is 0.89. Twenty-five percent of the observations have values below 10.20 and 25 percent have values above 11.44.

There is a substantial negative correlation (-0.63) between unemployment rate and resident occupational level. This is consistent with employment and job search theory. Higher-skilled workers suffer from less frequent periods of unemployment, lowering their overall unemployment rate, while lower-skilled workers tend to suffer from more frequent periods of unemployment. The economic restructuring of metropolitan economies, with demand for higher-skilled labor growing while demand for lower-skilled labor has stagnated or declined, has exacerbated the unemployment problems of lower-skilled workers (Kasarda, 1993).

There is a positive, but much weaker correlation (0.08) between occupational level and the local working rate. Because spatial job search theory suggests that higher occupation levels should result in less local working, this result is not expected. Higher-skilled workers conduct more spatially expansive job searches. The bivariate correlation may be masking any causal relationship. For example, occupational level is somewhat negatively correlated with proportion of residents who are black, which in turn is negatively correlated with local working. Thus higher occupational levels are associated with fewer black residents, which implies more local working. In fact, the results of Chapter 6 show that occupational level negatively affects the local working rate, consistent with theory.

Dissimilarity Between Occupations of Residents and Nearby Jobs

If the occupational mix of the neighborhood labor force differs substantially from the mix of jobs nearby, neighborhood residents seeking employment near the neighborhood might have difficulty finding suitable positions available. Thus, a high dissimilarity between the occupations of residents and the occupations of nearby jobs may result in lower local working as well as higher unemployment.

The dissimilarity index was calculated as follows:

$$\text{Occupational Dissimilarity} = \sum_{k=1}^{14} 1/2(\text{ABSOLUTE VALUE}(j_k/j - r_k/r)) \quad (11)$$

where

j = the total number of jobs within 2 miles of the residential quartersection;

j_k = the number of jobs in occupational class k within 2 miles of the residential quartersection, for k = 1 to 14;

r= the number of employed residents in the quartersection;

and

r_k= the number of residents of the quartersection employed in occupation k,
for k = 1 to 14.

The occupational dissimilarity index in the primary data set has a mean of 0.19, a median of 0.19 and a standard deviation of 0.06. With a possible range from zero to 1, the actual range is from 0.05 to 0.51. Twenty-five percent of the observations have an index below 0.15, and 25 percent have an index above 0.23.

Dissimilarity is positively but weakly correlated with unemployment.(0.21), generally consistent with job search theory. Given two neighborhoods with access to the same number of nearby jobs, the neighborhood where the occupations of nearby jobs are the most similar to those of the residents will have superior job opportunities. Dissimilarity is also negatively but weakly correlated with the local working rate (-0.16). The sign of this correlation is also as predicted. Similarity between job and resident occupations should allow for more residents to work near their neighborhood.

Race and Ethnicity

It is difficult to address urban phenomena in most metropolitan areas without considering race. Chicago is known to be one of the most racially segregated metropolitan areas in the United States. Due to the literature on the relationship between race and employment status, race must be considered as a key independent variable. This relationship is complicated by the fact that race is correlated both with employment status of neighborhood residents and with the nearby jobs-to-labor-force ratio. In the principal data set, however, the correlation between the proportion of residents who are black (proportion black) and unemployment (0.77) is much stronger than the correlation between the proportion black and the nearby jobs-to-labor-force ratio. (-0.29).

The literature clearly argues that race—specifically being black— is a strong determinant of employment status, after accounting for education, location, and other factors. Ihlanfeldt's work suggests that race may account for as much as one-half of the difference in unemployment rates among urban black and white youth (Ihlanfeldt, 1992). Kirschenman and Neckerman's interviews with Chicago area employers suggest discrimination and racial preference work primarily to the detriment of urban blacks (Kirschenman and Neckerman, 1991).

For the principal data set, the mean value of the proportion of residents who are black is 0.18, with a median of 0.01 and a standard deviation of 0.33. Values range from 0.00 to 1.00, and the twenty-fifth percentile is 0.00 while the seventy-fifth percentile is 0.13. The frequency distribution of this variable is essentially bimodal, with modes near 0.00 and 1.00. In addition to be positively related to the neighborhood unemployment rate, proportion black is negatively correlated (-0.37) to the local working rate.

For the principal data set, the mean value of the proportion of residents who are Hispanic is 0.08, the median is 0.03, and the standard deviation is 0.15. Values range from 0.00 to 0.95, with 25 percent of the observations having values of less than 0.01 and 25 percent having values above 0.08. Hispanics in Chicago are generally less segregated than blacks, and the frequency distribution of proportion Hispanic resembles a one-tailed, normal distribution. There is a much weaker correlation between proportion Hispanic and unemployment (0.17) compared to that between proportion black and unemployment (0.77). These relationships are consistent with the existing literature that suggests blacks suffer from higher unemployment rates than Hispanics. There is no substantial relationship between proportion Hispanic and local working rate, with a correlation coefficient of 0.0989.

Age

Variations among neighborhood unemployment rates may be partly due to variations in the age distributions of neighborhood residents. Unemployment problems are aggravated among adolescents and young adults, especially for black youth in central cities. Age is related to employment experience and occupation level. The proportion of a neighborhood's working-age persons, those 16 to 64, under 25 is expected to be a factor in determining the neighborhood unemployment rate. At the same time, youth might be expected to seek and attain employment near their neighborhoods at higher rates than older adults, due to decreased access to automobiles and to metropolitan-wide social and employment networks, as well as a greater overall reliance on neighborhood and family networks for job opportunities, as suggested by Wilson and O'Regan and Quigley (Wilson, 1987; O'Regan and Quigley, 1993).

For the principal data set, the mean of the proportion of working-age persons under age 25 is 0.19 with a median of 0.18 and a standard deviation of 0.06. The proportion under 25 ranges from 0.03 to 0.68 with 25 percent of the observations falling below 0.15 and 25 percent falling above 0.22. Many neighborhoods with large proportions of young residents are clustered on the west and south sides of the central city, where unemployment rates are high. There is significant correlation between age distribution and race/ethnicity, with the proportion under 25 being correlated both with proportion of residents who are black (0.3319) and Hispanic (0.3063). There is a substantial positive correlation (0.45) between the proportion under 25 and the neighborhood unemployment rate.

The correlation with the local working rate (0.04) is much weaker. While a positive correlation is consistent with the expectation that youth will tend to work nearer their own neighborhoods, the small magnitude of the correlation is somewhat surprising. This small magnitude is due in part to the higher unemployment rates of youth. If youth are less likely to be employed anywhere, then they are also less likely to be employed near their neighborhood. A stronger reason for the small correlation, however, is probably the correlation between proportion under 25 and proportion black (0.3319). Neighborhoods with more young working-age persons also tend to be more black, and black neighborhoods have lower local working rates. The ability of monovariate analysis to identify causal relationships is severely limited. The relationship between age and local working is examined again in the multiple regression results in Chapter 6.

Other Variables Included in the Analyses

Five other independent variables are included in model estimation in Chapter 6, including the proportion of the neighborhood labor force that is female, the occupational level of nearby jobs (calculated using the same median earning weights as in equation (10)), the rates of teenage and adult enrollment in schooling, and the distance of the neighborhood from the Chicago CBD. Table II shows that the average neighborhood has a labor force that is 46 percent female (0.46), with a median of 0.46 and a standard deviation of 0.05. Twenty-five percent of cases have more than 49 percent female labor forces with 25 percent having less than 43 percent female labor forces. Table III shows that this variable is positively correlated with unemployment and negatively correlated with local working. The latter is particularly unexpected. Women are expected to work closer to home to accommodate part-time work and greater responsibilities at home. The multivariate analysis in Chapter 6 finds that this variable affects unemployment negatively and local working positively, consistent with expectations.

Table II indicates that the average occupational level of nearby jobs for cases in the primary data set is 10.94, with a median of 10.94, and a standard deviation of 0.26. Comparing these figures to those describing the occupational level of residents reveals that the occupational level of nearby jobs is somewhat more uniform across space than the occupational level of neighborhoods. That is, nearby jobs are not as spatially segregated by occupation as residential neighborhoods. The occupational level of local jobs is positively correlated with unemployment (0.11), but not significantly correlated with local working. The former might be expected, as nearby jobs are not well matched to the skills of the unemployed, but the latter is not.

Table II also shows that the mean enrollment rate for teenagers aged 12 to 17 is 0.94 with a median of 0.96 and a standard deviation of 0.07. Twenty-five percent of the cases have enrollment rates below 91 percent. Teenage enrollment is negatively correlated with unemployment (-0.26), an expected result, and not significantly correlated with local working.

The mean adult enrollment rate (for all types of schooling) is 0.13 with a median of 0.12 and a standard deviation of 0.05. This variable is expected to affect local working greatly, as college students are expected to work near their residence more often. Adult enrollment is positively but mildly correlated with unemployment (0.06) and positively correlated with local working (0.15). The latter is as expected.

The mean distance from the CBD for cases in the primary data set is 15.55 miles, with a median of 14.04 and a standard deviation of 7.27 miles. Miles from the CBD is negatively correlated with unemployment (-0.40). This is expected as the Chicago urban structure is similar to many older cities in that wealthier, high-income households tend to live farther from the CBD. Miles from the CBD is not significantly related to local working. This is somewhat unexpected, as those closer to the CBD, but not within two miles of it, might be expected to be employed there, so that their local working rates would be lower. Indeed, the multivariate results of Chapter 6, which follow, fail to confirm any significant impact of miles from CBD on local working rates.

NOTES

1. For those residential quartersections within two miles of Lake Michigan, the catchment area is somewhat smaller.

2. Calculated from the 1990 CTPP Urban Element from Chicago. Area approximately includes northwestern Will county and most of Cook and DuPage counties. See Figure 3 for a map of the included area.

3. The central business district is defined here as running from Van Buren (400 S.) on the south; Chicago Avenue (800 N.) on the north; Halsted (800 W.) on the west; and Lake Michigan on the east.

4. Based on unit counts and addresses obtained from the Chicago Housing Authority, April, 1995.

5. The low- and moderate-skilled labor force is equal to the labor force excluding those employed in executive/managerial; professional specialty; technician and related support; and precision production, craft and repair occupations.

6. Spatial plots of variables each utilize four ranges for the variable of interest. In general, if the mean is larger than the standard deviation, the first range is from zero to the mean minus the standard deviation, and subsequent levels are in standard deviation increments. For variables where the mean is smaller than the standard deviation, the first range is from zero to the mean, and subsequent levels are in standard deviation increments. For proportion black, a bipolar variable, standard deviation increments are not used.

7. High-skilled jobs include those in the categories of executive and managerial; professional; technical and related support; and precision production and craft.

Estimating the Models and Interpreting the Results

ESTIMATING A MODEL OF NEIGHBORHOOD UNEMPLOYMENT

In order to identify the contributions of the various independent variables to neighborhood unemployment, simple monovariate correlations are not sufficient. The data set described in the previous chapter can be used to estimate a multivariate model that explains neighborhood unemployment rates, including the effect of nearby jobs. Following these results, a similar model is used to explain neighborhood local working rates.

The Basic Linear Model

Equation (8) of Chapter 4 can be estimated from the principal data set. Results are presented in Table III. Most independent variables are significant at below $p = 0.01$. Teenage enrollment in school and the proportion of the labor force who are female are significant at 0.05 or below; and the proportion Hispanic is not significant at any reasonable confidence level. Table IV gives the change in neighborhood unemployment rate associated with a change in the independent variable of one standard deviation.

While the effect of an increase in the jobs-to-labor-force ratio is statistically significant, it is relatively small, especially when compared to the effect of changes in some of the other variables. A one standard deviation increase in jobs-to-labor-force from 0.87 to 1.39 results in a

0.3 percentage point decrease in unemployment rate, or about 4 percent of the mean unemployment rate of 7 percent. Substantially larger effects result from changes in two other variables. A one standard deviation increase in proportion black from 0.18 to 0.51, for example, increases the neighborhood unemployment rate by 3.6 percentage points, and a one standard deviation increase in the resident occupational level index from 10.8 to 11.69 decreases unemployment by 1.9 percentage points.

While the impact of the jobs-to-labor-force ratio, by itself is not very large, the occupational similarity between residents and nearby jobs as well as the overall occupational level of the jobs also affect unemployment. It appears that the magnitude of the nearby jobs-to-labor-force ratio alone is less important than the occupational similarity and the occupation level of the jobs. A standard deviation decrease in the occupational dissimilarity between residents and nearby jobs, for example, reduces predicted unemployment by 0.7 percentage points. A standard deviation decrease in the occupation level of nearby jobs reduces predicted unemployment by 0.5 percentage points. So the nature of nearby jobs, especially in reference to the particular occupational mix of the neighborhood, appears to be more important than the aggregate number of nearby jobs in determining unemployment.

Notwithstanding the overall impact of the three job-related variables, the largest determinant of neighborhood unemployment is the proportion of residents who are black. This is not surprising given the literature on black employment problems. The second most important independent variable is the occupational level of residents. The importance of this labor supply characteristic is also expected. Occupational level is a strong indicator of overall education and training or human capital. Other independent variables for which a standard deviation change results in at least a 0.5 percentage point change in the predicted unemployment rate include miles from the central business district (CBD) and the proportion of working-age persons under 25. A seven mile increase in distance from the CBD results in a decrease in predicted unemployment of 1 percentage point. A six percentage point decrease in the proportion of working-age persons under 25 results in an increase in predicted unemployment of 0.7 percentage points.

Table III. Results of OLS Regression:
Neighborhood Unemployment Rate Estimation, Equation (8)

Dependent Variable: Unemployment Rate (0.00 to 1.00)

Multiple R	0.86540
R Square	0.74892
Adjusted R Square	0.74723
Standard Error	0.03258

Analysis of Variance

	Degrees of Freedom	Sum of Squares	Mean Square
Regression	11	5.16931	0.46994
Residual	1633	1.73304	0.00106

N = 1,645

F = 442.81048   Significance F = 0.0000

Variables in the Equation

Variable	*Coefficient*	*Standard Error*	*t-Statistic*
Jobs-to-Labor-Force	-0.005649	0.001645	- 3.435 ***
Resident Occupational Level	-0.021022	0.001472	-14.286 ***
Occupational Dissimilarity	0.113635	0.014140	8.037 ***
Proportion Black	0.107706	0.004000	26.878 ***
Proportion Hispanic	0.010999	0.007838	1.403
Female Labor Force	-0.052666	0.021051	-2.502 **
Working-Age Persons Under 25	0.111442	0.019960	5.583 ***
Teenage Enrollment	-0.027669	0.012746	-2.171 *
Adult Enrollment	-0.060549	0.020325	-2.979 ***
Job Occupational Level	0.018408	0.003448	5.338 ***
Miles from CBD	-0.001337	1.2918E-04	-10.347 ***
(Constant)	0.113970	0.039635	2.876 ***

*** significant at below 0.01
** significant at below 0.05
* significant at below 0.10

Table IV. Change in Unemployment Due to Standard Deviation Change in Independent Variable

Independent Variable	Mean	Standard Deviation	Coefficient	Coefficient x Standard Deviation
Jobs-to-Labor-Force Ratio	0.87	0.52	-0.005649	-0.003
Resident Occupational Level	10.80	0.89	-0.021022	-0.019
Occupational Dissimilarity	0.19	0.06	0.113635	0.007
Proportion Black	0.18	0.33	0.107706	0.036
Female Labor Force	0.46	0.05	-0.052666	-0.003
Working Age Persons Under 25	0.19	0.06	0.111442	0.007
Teenage Enrollment	0.94	0.07	-0.027669	-0.002
Adult Enrollment	0.13	0.05	-0.060549	-0.003
Job Occupational Level	10.94	0.26	0.018408	0.005
Miles from CBD	15.55	7.27	-0.001337	-0.010

The miles from CBD variable could be capturing one or both of two potential effects. First, increasing distance from the CBD may signal greater proximity to suburban job centers—a proximity beyond a two-mile radius. Being closer to these suburban job centers may result in better access to the jobs there and therefore lower unemployment. Second, miles from the CBD may be capturing differences in education, skills and access to job networks not measured by other variables such as occupational level, race and enrollment rates. The occupational level index, for example, does not distinguish between "professionals" who are physicians and those who are teachers. Middle management is also not distinguished from upper management. Those living in wealthier neighborhoods, farther from the CBD, are likely to suffer less frequent bouts of unemployment. Thus, miles from CBD may help measure the unmeasured skill and social network characteristics that can be important to determining unemployment. This variable will be discussed more below.

The proportion of the neighborhood labor force that is female also has a negative effect on unemployment. This finding is consistent with the high rates of unemployment among young, predominantly minority men, which is the subject of a large amount of the literature. As expected, teenage and adult enrollment rates are significant and negative on unemployment, although their impacts are relatively modest. Higher teenage enrollment rates should be correlated with better educational achievement among young adults and, perhaps, with better educational processes at local schools. Higher adult enrollment rates indicate a population that is pursuing education, raising their overall skill level.

The Need to Consider Nonlinear Jobs-to-Labor-Force Terms

Simpson finds that the use of a quadratic term on his job-to-resident ratio adds significant explanatory power to his model (Simpson, 1992). He suggests that areas with very high job densities may attract workers from outside the area, thereby increasing competition for these jobs and hurting the employment prospects of residents. Thus, additional jobs actually leads to higher unemployment at high job-to-labor-force levels.

There is, however, an additional source of potential nonlinearity in the jobs-to-labor-force versus unemployment specifications: a simultaneity problem between jobs-to-labor-force and unemployment of the type often encountered in this type of research (Ihlanfeldt, 1992; and Holzer, Ihlanfeldt and Sjoquist, 1994). Some households may prefer less commercial, more residential areas for their places of residence. They may be seeking to avoid pollution, traffic congestion, and other negative externalities caused even by relatively modest numbers of employers and jobs. If this is true, these households will prefer to live in areas near few jobs and, once residing in such areas, may act to preserve the residential character of these communities. Areas with low jobs-to-labor-force, therefore, may be desirable residential areas for this segment of the population. If these adults are rarely unemployed, these neighborhoods may exhibit low unemployment rates and low jobs-to-labor-force ratios at the same time.

This type of effect may also explain Simpson's quadratic estimation, although he may have encountered a different version of the phenomena. Neighborhoods with very high jobs-to-labor-force ratios may not be very desirable places to live, and may be left to lower-income unemployment-prone households. Very high nearby job levels, such as near O'Hare Airport, may be associated with negative pollution or congestion externalities. Thus, the preference for lower-jobs-to-labor-force ratios might affect the linearity of the unemployment specification at both low and high jobs-to-labor-force levels. Some households may choose neighborhoods with very low jobs-to-labor-force levels over those with moderate levels, for example, while others prefer moderate jobs-to-labor-force areas to those with very high ratios. Such a scenario could result in higher unemployment as jobs-to-labor-force increases at either low or very high jobs-to-labor-force levels. These negative effects on unemployment are not due to changes in job access, but due to residential location preferences of households. This phenomenon, together with the expectation that higher jobs-to-labor-force ratios should increase employment prospects for neighborhood residents seeking work, suggests respecifying equation (8) by adding quadratic and cubic jobs-to-labor-force terms.

In order to identify potential nonlinearities, consider the following revision to equation (8):

$$u = \alpha + \beta_1(J_{LM}/R_{LM}) + \beta_2(J_{LM}/R_{LM})^2 + \beta_3(J_{LM}/R_{LM})^3 + \gamma(x_1, \ldots x_k) \quad (12)$$

Quadratic and cubic jobs-to-labor-force terms are added to allow for the fact that low levels of jobs-to-labor-force may be associated with more residentially desirable neighborhoods, preferred by more employable households. At the same time, very high levels of jobs-to-labor force (e.g., near O'Hare airport) may be associated with severe negative externalities that make nearby neighborhoods less desirable. Thus, β_1 is predicted to be positive, β_2 is predicted to be negative, and β_3 is predicted to be positive.

Table V presents the results of estimating equation (12) with the principal data set. Again, all variables except proportion Hispanic are significant, although the linear jobs-to-labor- force variable is now significant only at the $p = 0.07$ level. Table VI indicates the change in unemployment associated with a change of one standard deviation in each significant independent variable. For most variables the results are essentially the same as those in Table IV. So this more sophisticated specification does not significantly affect the results for the other variables.

The signs of the coefficients of the jobs-to-labor-force terms are as expected, consistent with the explanation of residential location effects at high and low jobs-to-labor-force levels. The interpretation of the magnitudes of the jobs-to-labor-force terms is more complex than for the simpler results in Tables III and IV. For example, a 0.5 increase in the jobs-to-labor-force ratio from 0.1 to 0.6 results in an increase in predicted unemployment of 0.5 percentage points, but an increase from 1.0 to 1.5 results in a decrease in predicted unemployment of -0.5 percentage points.

The addition of quadratic and cubic terms means that the impact of a change in jobs-to-labor-force varies depending on the base level of the variable. Figure 5 plots the estimated equation (12), giving predicted unemployment versus the jobs-to-labor-force ratio. Comparing the peak at 0.6836 to the median jobs-to-labor-force ratio of 0.75 indicates that, for almost one-half of the observations, a marginal increase in the jobs-to-labor-force ratio results in an increase in predicted unemployment. The slope of the predicted unemployment function, $u(J_{LM}/R_{LM})$, over J_{LM}/R_{LM} can be obtained by differentiating as follows:

$$du(J_{LM}/R_{LM})/d(J_{LM}/R_{LM}) = \beta_1 + 2\beta_2(J_{LM}/R_{LM}) + 3\beta_3(J_{LM}/R_{LM})^2 \qquad (13)$$

Table V. Results of OLS Regression: Neighborhood Unemployment Rate Estimation, Equation (12) (Cubic Form)

Dependent Variable: Unemployment Rate (0.00 to 1.00)

Multiple R	0.86628
R Square	0.75044
Adjusted R Square	0.74845
Standard Error	0.03250

Analysis of Variance

	Degrees of Freedom	Sum of Squares	Mean Square
Regression	13	5.17980	0.39845
Residual	1631	1.72255	0.00106

N= 1,645

F = 377.26919   Significance F = 0.0000

Variables in the Equation

Variable	Coefficient	Standard Error	t-Statistic
Jobs-to-Labor-force	0.015527	0.008353	1.859 *
(Jobs-to-Labor-force)^2	-0.014013	0.004814	-2.911 ***
(Jobs-to-Labor-force)^3	0.002150	6.9428E-04	3.096 ***
Resident Occupational Level	-0.020972	0.001468	-14.284 ***
Occupational Dissimilarity	0.113311	0.014106	8.033 ***
Proportion Black	0.109047	0.004123	26.450 ***
Proportion Hispanic	0.012028	0.007851	1.532
Female Labor Force	-0.051793	0.021002	-2.466 **
Working-Age Persons Under 25	0.111537	0.019920	5.599 ***
Teenage Enrollment	-0.027185	0.012725	-2.136 **
Adult Enrollment	-0.062660	0.020287	-3.089 ***
Job Occupational Level	0.019556	0.003479	5.622 ***
Miles from CBD	-0.001298	1.2996E-04	-9.987 ***
(Constant)	0.091555	0.040179	2.279 **

*** significant at 0.01
** significant at 0.05
* significant at 0.10

Table VI. Change in Unemployment Due to Standard Deviation Change in Independent Variable for Equation (12) (Cubic Form)

Independent Variable	Mean	Standard Deviation	Coefficient	Coefficient x Standard Deviation
Jobs-to-Labor-Force	0.87	0.52	0.015527	0.008
(Jobs-to-Labor-Force)^2	*	*	-0.014013	-0.016
(Jobs-to-Labor-Force)^3	*	*	0.002150	0.004
Resident Occupational Level	10.80	1.04	-0.020972	-0.022
Occupational Dissimilarity	0.19	0.06	0.113311	0.007
Proportion Black	0.18	0.33	0.109047	0.036
Female Labor Force	0.46	0.05	-0.051793	-0.003
Working Age Persons Under 25	0.19	0.06	0.111537	0.007
Teenage Enrollment	0.94	0.07	-0.027185	-0.002
Adult Enrollment	0.13	0.05	-0.062660	-0.003
Job Occupational Level	10.94	0.26	0.019556	0.005
Miles from CBD	15.55	7.27	-0.001298	-0.009

* Note: Values correspond to Jobs-to-Labor-Force Mean= 0.87; Standard Deviation = 0.52

Figure 5. Plot of Predicted Unemployment Versus Jobs-To-Labor-Force Ratio.

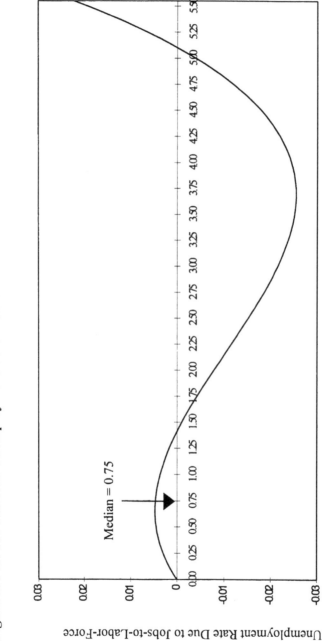

Solving for the zeros of this quadratic equation reveals the values of J_{LM}/R_{LM} where unemployment increases with increasing J_{LM}/R_{LM} and the values where unemployment declines with J_{LM}/R_{LM}. The solution gives:

$$J_{LM}/R_{LM} = \{-2\beta_2 +/- [(2\beta_2)^2 - 4 (3\beta_3)(\beta_1)]^{0.5}\}/ 2(3\beta_3) = 0.6836; 3.8735 \quad (14)$$

These two points are the peak and trough, respectively, of the cubic plot of Figure 5. So, for jobs-to-labor-force values less than 0.6836, increases in jobs-to-labor-force result in higher predicted unemployment. For values between 0.6836 and 3.8735, increases result in lower unemployment. For values above 3.8735, increases again result in higher neighborhood unemployment. An increase in jobs-to-labor-force from its mean of 0.87 to 1.39, a standard deviation increase, for example, results in a predicted change in unemployment equal to:

$$
\begin{aligned}
\Delta U \quad &= \quad (0.015527)*(1.39\text{-}0.87) + (-0.014013)*(1.39^2\text{-}0.87^2) + \\
&\quad (0.002150)*(1.39^3\text{-}0.87^3) \\
&= \quad 0.0081 - 0.0165 + 0.0044 = -0.004 \quad\quad (15)
\end{aligned}
$$

But for a similar magnitude increase from 1.39 to 1.91, the results are:

$$
\begin{aligned}
\Delta U \quad &= \quad (0.015527)*(1.91\text{-}1.39) + (-0.014013)*(1.91^2\text{-}1.39^2) + \\
&\quad (0.002150)*(1.91^3\text{-}1.39^3) \\
&= \quad 0.0081 - 0.0240 + 0.0089 = -0.007 \quad\quad (16)
\end{aligned}
$$

In fact, the maximum slope for $u(J_{LM}/R_{LM})$ occurs at $J_{LM}/R_{LM} = 2.17$ (a relatively high value of jobs-to-labor-force), around which a standard deviation increase (from 1.91 to 2.43) yields a 0.8 percentage point decrease in predicted unemployment.

These results are not consistent with those of Simpson, who found that unemployment decreased over a larger range of lower jobs-to-labor-force values and then increased only at higher levels. In addition, these results suggest a problem with Simpson's explanation of his results. He argues that areas with higher jobs-to-labor-force levels draw workers from a larger radius and so local residents have more difficulty accessing local jobs. It may be that he was actually detecting the phenomenon I suggest above, where households prefer not to live in areas with very high jobs-to-labor-force levels. This explains the second zero of the equation (14), shown as the trough in Figure 5.

The results for the effects of J_{LM}/R_{LM}, $(J_{LM}/R_{LM})^2$, and $(J_{LM}/R_{LM})^3$ suggest that, after adjusting for preferences of more employable households for low jobs-to-labor-force areas, the effect of nearby jobs may be significantly greater than suggested by the results in Table IV. Thus the effect of a standard deviation increase in jobs-to-labor-force, net of the residential preference effect, may exceed the 0.8 percentage point decrease in unemployment identified above.

Stratifying the Principal Data Set on Jobs-to-Labor-Force

The results above suggest that an unmeasured factor other than changes in access to jobs may be accounting for lower unemployment rates in areas with lower jobs-to-labor-force ratios. If, as suggested, preferences for low jobs-to-labor-force residential locations by more employable households are affecting the results, then the magnitude of the jobs-to-labor-force coefficients in the estimation of equation (8) may be biased downward. At the same time, however, it appears unlikely that correcting for any such bias would dramatically increase the effect if jobs-to-labor-force on neighborhood unemployment rates.

Besides utilizing the cubic functional form, another way to attempt to isolate the effect of unemployment on jobs-to-labor-force is to stratify the principal data by separating low jobs-to-labor-force neighborhoods from the other quartersections. The mode jobs-to-labor-force ratio is 0.55. The full principal data set can be stratified into two subsets, one with jobs-to-labor-force less than or equal to 0.55 and one with values greater than 0.55. Of the cases in the working data, 387 (24 percent) have values less than or equal to 0.55, leaving 1,258 cases in the upper subset. Table VII gives the OLS results for estimating equation (8), the linear model, with the upper subset only. Comparing these results with those of Table III reveals that the coefficient on jobs-to-labor-force is larger in magnitude for the upper subset, increasing from -0.005649 to -0.008919.

Given that the entire primary data set fits a cubic functional form, the upper subset should still be expected to exhibit some nonlinearity in jobs-to-labor-force, particularly a quadratic form. Estimating with a quadratic form yields the results in Table VIII. The results are relatively consistent with those in Tables V and VI.

Table VII. Results of OLS Regression: Neighborhood Unemployment Rate Estimation, Equation (8) With Only Cases Where $J_{LM}/R_{LM} > 0.55$

Dependent Variable: Unemployment Rate (0.00 to 1.00)

Multiple R	0.85888
R Square	0.73767
Adjusted R Square	0.73536
Standard Error	0.03032

Analysis of Variance

	Degrees of Freedom	Sum of Squares	Mean Square
Regression	11	3.22021	0.29275
Residual	1246	1.14516	0.00092

N= 1,258

F = 318.52616   Significance F = 0.0000

Variables in the Equation			
Variable	Coefficient	Standard Error	t-Statistic
Jobs-to-Labor-Force	-0.008919	0.001728	-5.163 ***
Resident Occupational Level	-0.019068	0.001633	-11.680 ***
Occupational Dissimilarity	0.111164	0.015606	7.123 ***
Proportion Black	0.129740	0.005009	25.900 ***
Proportion Hispanic	0.021756	0.008408	2.588 ***
Female Labor Force	-0.039416	0.021825	-1.806 *
Working-Age Persons Under 25	0.081399	0.02130	3.820 ***
Teenage Enrollment	-0.022397	0.013227	-1.693 *
Adult Enrollment	-0.058542	0.021141	-2.769 ***
Job Occupational Level	0.014453	0.003720	3.886 ***
Miles from CBD	-0.001127	1.3586E-04	-8.296 ***
(Constant)	0.129192	0.040748	3.170 ***

*** significant at 0.01

** significant at 0.05

* significant at 0.10

Table VIII. Results of OLS Regression: Neighborhood Unemployment Rate Estimating Equation (8) Plus Quadratic Term Only Cases With $J_{LM}/R_{LM} > 0.55$

Dependent Variable: Unemployment Rate (0.00 to 1.00)

Multiple R	0.86053
R Square	0.74050
Adjusted R Square	0.73800
Standard Error	0.03016

Analysis of Variance

	Degrees of Freedom	Sum of Squares	Mean Square
Regression	12	3.23257	0.26938
Residual	1245	1.13280	0.00091

N= 1,258
F = 296.06249   Significance F = 0.0000

Variables in Equation

Variable	Coefficient	Standard Error	t-Statistic
Jobs-to-Labor-force	-0.024105	0.004465	-5.399 ***
$(\text{Jobs-to-Labor-force})^2$	0.003950	0.001072	3.685 ***
Resident Occupational Level	-0.019220	0.001625	-11.828 ***
Occupational Dissimilarity	0.111327	0.015528	7.169 ***
Proportion Black	0.128007	0.005006	25.569 ***
Proportion Hispanic	0.019766	0.008383	2.358 **
Female Labor Force	-0.039788	0.021715	-1.832 *
Working-Age Persons Under 25	0.082040	0.021201	3.870 ***
Teenage Enrollment	-0.023786	0.013166	-1.807 *
Adult Enrollment	-0.058380	0.021035	-2.775 ***
Job Occupational Level	0.016761	0.003754	4.465 ***
Miles from CBD	-0.001097	1.3543E-04	-8.097 ***
(Constant)	0.117070	0.040677	2.878 ***

*** significant at 0.01
** significant at 0.05
* significant at 0.10

The results in Table VIII suggest that an increase in jobs-to-labor-force from 0.87 to 1.39, a standard deviation for the entire principal data set, results in a decrease in predicted unemployment of 0.8 percentage points. This is a larger effect but is still not nearly as large as the race and skill effects. An increase from 3.00, a very high jobs-to-labor-force value, to 3.52 results in an increase in predicted unemployment of only 0.1 percentage points.

The Effect of Nearby Jobs on Neighborhood Unemployment

The results above suggest that higher jobs-to-labor-force ratios reduce predicted neighborhood unemployment, but the size of the effect is modest. A standard deviation increase in jobs-to-labor-force from 0.87 to 1.39 results in a decrease in predicted unemployment on the order of 0.8 percentage points, though the magnitude of this result may be biased somewhat downward.

The jobs-to-labor-force ratio, however, is not the only variable that describes nearby labor demand. Other characteristics of nearby jobs are contributors to the variation in neighborhood unemployment rates. One characteristic is the dissimilarity between the occupations of the employed residents of the quartersection and the jobs within two miles of the quartersection. Tables IV and VI suggest that an increase in occupational dissimilarity of 0.06, or one standard deviation, will increase predicted unemployment by 0.7 percentage points. The average occupational level of the nearby jobs also affects neighborhood unemployment. Tables IV and VI show that an increase in average occupational level of nearby jobs of 0.26, one standard deviation, will increase predicted unemployment by 0.5 percentage points. Thus, the impact of nearby jobs on unemployment is considerably larger when accounting for the occupational mix of jobs vis-a-vis residents and the average occupational level of the jobs. Cumulatively, standard deviation changes in predicted unemployment due to standard deviation changes in all three variables (jobs-to-labor-force, occupational dissimilarity, and job occupation level) can amount to as much as 2 percentage points, equal to one-third of the standard deviation for unemployment. This is without considering the effects net of any preference of more employable households for low jobs-to-labor-force areas, which would result in a larger cumulative effect. The results in this chapter suggest that job creation and retention in and around the neighborhood reduce unemployment modestly. They also indicate that nearby jobs will reduce neighborhood unemployment more if the particular occupational mix of nearby jobs is not too different from the occupational mix of residents.

Finally, the findings here suggest that the impact of neighborhood jobs will be greater if the occupational levels of the jobs are not too high. Low-skill jobs, such as handlers and laborers, will reduce unemployment more than moderate-skill jobs such as administrative support, and moderate-skill jobs will reduce unemployment more than high-skill jobs. This is due to the low skill levels of many of the unemployed.

The Effects of Other Neighborhood Characteristics on Neighborhood Unemployment

Besides variables that describe the jobs near a neighborhood with respect to neighborhood residents, the results in Tables III through VIII show that a number of resident characteristics affect neighborhood unemployment. Two variables, proportion black and resident occupational level, have a major impact on neighborhood unemployment, with standard deviation changes in the independent variables resulting in changes in the predicted unemployment rate of more than two percentage points. Standard deviation changes in two other independent variables, including miles from CBD and the proportion of working-age residents under age 25, result in changes in unemployment between 0.5 and 0.9 percentage points. Finally, standard deviation changes in adult enrollment, teenage enrollment, and the proportion of the labor force that is female have small but significant impacts on unemployment. The effects of variables in the first two categories merit further discussion.

Race and Ethnicity

The results in Tables III through VIII reveal that the largest single contributor to neighborhood unemployment is the proportion of residents who are black. A one standard deviation increase in proportion black (0.33) results in a 3.6 percentage point increase in predicted unemployment. Thus, going from a completely nonblack neighborhood to an all-black neighborhood increases predicted unemployment by 10.8 percentage points. These findings are not surprising given the literature on black employment problems and a familiarity with these problems in Chicago.

Black unemployment problems may arise not only due to employment discrimination, although this is a likely contributor here, but also due to potentially unmeasured differences in educational and occupational experiences. Urban blacks suffer from inferior access to basic education (Orfield, 1992). Resulting poor educational achievement combined with inferior access to job opportunities helps reinforce a dual employment market (Cain, 1976). Unlike the proportion black variable, the proportion of residents who are Hispanic is generally not significant in the results of Tables III through VI. In regressing with the entire primary data set, proportion Hispanic comes in positive but becomes significant only at the $p = 0.17$ and $p = 0.13$ levels. The coefficient on proportion Hispanic is only slightly more than one-tenth the magnitude of the coefficient on proportion black, and the standard error is larger. The relative insignificance of this variable and the relatively weak impact of Hispanic status vis-à-vis black status is consistent with the literature on black and Hispanic employment prospects. Blacks suffer more from employment discrimination, and in some cases, employers appear to prefer Hispanic workers even over white workers (Kirschenman and Neckerman, 1991).

Occupational Level of Residents

The second most important independent variable is the resident occupational level. A one standard deviation increase in resident occupational level results in a decrease in predicted unemployment of between 1.9 and 2.2 percentage points, depending on the specification. The negative sign and significance of this labor supply characteristic are expected. Resident occupational level is a strong indicator of overall education and training or human capital. The findings on this variable, as well as the occupational dissimilarity variable, corroborate Kasarda's and Wilson's theses that the low-level occupations and skills of many urban residents explain a good deal of employment status outcomes in an increasingly technological economy (Kasarda, 1993; Wilson, 1987). The strength of this variable vis-à-vis the jobs-to-labor-force variable suggests that skill mismatch may explain more of the urban neighborhood unemployment problem than spatial mismatch, although both are significant.

Miles from the Central Business District

Because the central business district (CBD) contains many jobs and few residents, proximity to it might be expected to reduce unemployment. The regression results suggest otherwise. Miles from CBD is strongly significant but has a substantial, negative effect on neighborhood unemployment, with an increase of slightly more than seven miles from the CBD decreasing unemployment by one percentage point. This result may reflect the spatial constraints and poor access of central city and near-in neighborhoods to distant newer employment opportunities in the outer parts of the metropolitan area. This poor access, in turn, is caused by the decentralization of employment opportunities, housing discrimination in job-rich suburbs, inadequate transportation to newly developed job centers, and social isolation.

Recent research suggests that the Chicago metropolitan area may actually exhibit characteristics of a polycentric urban model, where multiple job centers, in addition to the CBD, exist (McDonald and Prather, 1994). Given such a model, proximity to secondary employment centers might affect neighborhood unemployment rates. McDonald and Prather identify newer, secondary job centers, including one near O'Hare Airport, in the northwest part of the study area, and one in central DuPage County, in the western part of the study area. Proximity to these centers, beyond that captured for neighborhoods within two miles of these two sites, may reduce neighborhood unemployment rates. Table IX provides OLS results for the cubic specification of Table V but with two additional independent variables, measuring miles from O'Hare and Central DuPage, respectively, in the regression. The results show that distance from O'Hare, the larger of the two secondary job centers, is significant and positive on unemployment, but only at a $p = 0.06$ level and only with a relatively small coefficient. An increase of seven miles from O'Hare results in an increase in predicted unemployment of 0.3 percentage points. This result suggests that, if the effect of distance from the CBD on unemployment is due to inferior access to suburban jobs, the access

Table IX. Results of OLS Regression: Neighborhood Unemployment Rate Estimation, Equation (12) (Cubic Form) Including Secondary Job Center Variables

Dependent Variable: Unemployment Rate (0.00 to 1.00)

Multiple R	0.86727
R Square	0.75216
Adjusted R Square	0.74988
Standard Error	0.03241

Analysis of Variance

	Degrees of Freedom	Sum of Squares	Mean Square
Regression	15	5.19166	0.34611
Residual	1629	1.71069	0.00105

N = 1,645

F = 329.58199   Significance F = 0.0000

Variables in the Equation

Variable	Coefficient	Standard Error	t-Statistic
Jobs-to-Labor-force	0.017216	0.00838	2.052 **
(Jobs-to-Labor-force)^2	-0.013643	0.004833	-2.823 ***
(Jobs-to-Labor-force)^3	0.002065	6.9565E-04	2.968 ***
Resident Occupational Level	-0.021021	0.001465	-14.353 ***
Occupational Dissimilarity	0.110404	0.014185	7.783 ***
Proportion Black	0.104061	0.004372	23.803 ***
Proportion Hispanic	0.009372	0.007908	1.185
Female Labor Force	-0.050174	0.020969	-2.393 **
Working-Age Persons Under 25	0.116671	0.019953	5.847 ***
Teenage Enrollment	-0.025188	0.012704	-1.983 **
Adult Enrollment	-0.068892	0.020350	-3.385 ***
Job Occupation Level	0.023471	0.003660	6.413 ***
Miles from CBD	-0.001537	1.5456E-04	-9.947 ***
Miles from DuPage	1.00694E-04	2.3541E-04	0.428
Miles from O'Hare	3.74319E-04	1.9861E-04	1.885 *
(Constant)	0.042831	0.042686	1.003

*** significant at 0.01

** significant at 0.05

* significant at 0.10

problem is too complex to be explained by the distance to one or two secondary job centers.

Proportion of Working-Age Persons Under 25

A standard deviation increase of 0.6 percentage points in the proportion of working-age persons who are under 25 results in an increase in predicted unemployment of 0.7 percentage points. This is generally consistent with theory and the literature. Young adults and teenagers, especially in minority communities, suffer from high unemployment rates. Age, like occupational level and enrollment rates, is a measure of human capital. In addition, youth may have inferior access to automobile transportation, may have developed less extensive occupational contacts, and may have penetrated fewer job networks.

ESTIMATING THE LOCAL WORKING RATE

In neighborhoods with very low jobs-to-labor-force ratios, the proportion of residents in the labor force that will be employed near the neighborhood will be very low. They will find very few job opportunities there. As jobs become available nearby, those preferring to work nearby or those who are highly transportation- or information-constrained—especially part-time workers and youth—will quickly seek out and attempt to obtain nearby jobs. As the jobs-to-labor-force ratio increases, there will be fewer residents who still seek and have not already attained nearby work, so the local working rate will increase more slowly and then reach a plateau. However, it is likely that, in areas with small numbers of residents and extremely large jobs-to-labor-force ratios, a relatively large proportion of residents will work nearby. This is because some workers at nearby jobs may move into the neighborhood to be closer to their work places. Even if only a small proportion of a large number of nearby workers move into a sparsely populated neighborhood, the neighborhood's local working rate may increase significantly.

Applying the Cubic Specification to Estimating the Local Working Rate

The above scenario can be approximately modeled by using a cubic specification similar to that used in modeling the neighborhood unemployment rate. The cubic functional form of equation (12) can be used as a model for the local working rate, w:

$$w = \alpha + \beta_1(J_{LM}/R_{LM}) + \beta_2(J_{LM}/R_{LM})^2 + \beta_3(J_{LM}/R_{LM})^3 + \gamma(x_1, \ldots x_k) \quad (17)$$

Using the same set of neighborhood characteristics, $x_1, \ldots x_k$, as presented in Table V gives the OLS results shown in Table X, with impacts of standard deviation changes in significant variables on w shown in Table XI. The explanatory power of this regression, while substantial, is noticeably lower than the estimation of unemployment in Table V. The R^2 here is slightly less than 0.40, while the R^2 in Table V is just over 0.75. This is expected as more unmeasured explanatory variables are missing here, including the particular hiring practices of nearby firms, transportation characteristics of the nearby area (which could affect the barriers to working farther from the neighborhood), the development of local hiring networks between residents and firms, etc. While some of these factors may affect unemployment as well as local working, they are more important to local working. Labor supply characteristics, including race and skill, are more important in determining whether people attain work, than in determining where they work. The latter question is more complex and is less explainable with the data here.

Table XI shows that a one standard deviation increase in jobs-to-labor-force from 0.87 to 1.39 results in an increase in w of 0.130 - 0.117 + 0.023= 0.036, or 3.6 percentage points. Given a mean local working rate of 15 percent, this is a substantial effect. Solving for the zeros of equation (13) again, except with the coefficient values from Table XI, gives the following zeros:

$$J_{LM}/R_{LM} = \{-2\beta_2 +/- [(2\beta_2)^2 - 4(3\beta_3)(\beta_1)]^{0.5}\}/2(3\beta_3) = 1.86; 3.90 \quad (18)$$

These zeros are both relatively large. (The 75th percentile for jobs-to-labor-force is 1.05.) This means that increased jobs-to-labor-force results in increased predicted local working for most neighborhoods, but that as jobs-to-labor-force becomes large, local working increases less, reaches a plateau and then begins to decline slightly as jobs-to-labor-force exceeds 1.86. Then, as jobs-to-labor-force exceeds 3.90, local working begins to increase again.

Table X. Results of OLS Regression: Neighborhood Local Working Rate (W) Estimation, Equation (15) (Cubic Form)

Dependent Variable: w (0.00 to 1.00)

Multiple R	0.62903
R Square	0.39568
Adjusted R Square	0.39086
Standard Error	0.06070

Analysis of Variance

	Degrees of Freedom	Sum of Squares	Mean Square
Regression	13	3.93518	0.30271
Residual	1631	6.01019	0.00368

N = 1,645

F = 82.14613   Significance F = 0.0000

Variables in the Equation

Variable	Coefficient	Standard Error	t-Statistic
Jobs-to-Labor-force	0.250226	0.015603	16.037 ***
(Jobs-to-Labor-force)^2	-0.099261	0.008992	-11.038 ***
(Jobs-to-Labor-force)^3	0.011477	0.001297	8.850 ***
Resident Occupational Level	-0.006444	0.002742	-2.350 **
Occupational Dissimilarity	-0.097947	0.026349	-3.717 ***
Proportion Black	-0.058042	0.007701	-7.537 ***
Proportion Hispanic	0.020185	0.014666	1.376
Female Labor Force	0.070181	0.039231	1.789 *
Working-Age Residents Under 25	0.114681	0.03720	3.082 ***
Teenage Enrollment	-0.018254	0.023769	-0.768
Adult Enrollment	0.254453	0.037894	6.715 ***
Job Occupational Level	0.002252	0.006498	0.347
Miles from CBD	2.00369E-04	2.4276E-0	0.825
(Constant)	0.016878	0.075051	0.225

*** significance at 0.01

** significance at 0.05

* significance at 0.10

Table XI. Change in Local Working Rate (W) Due to Standard Deviation Change in Independent Variable for Equation (15) (Cubic Form)

Independent Variable	Mean	Standard Deviation	Coefficient	Coefficient x Standard Deviation
Jobs-to-Labor-Force	0.87	0.52	0.250226	0.130
(Jobs-to-Labor-Force)^2	*	*	-0.099261	-0.117
(Jobs-to-Labor-Force)^3	*	*	0.011477	0.023
Resident Occupational Level	10.8	0.89	-0.006444	-0.006
Occupational Dissimilarity	0.19	0.06	-0.097947	-0.006
Proportion Black	0.18	0.33	-0.058042	-0.019
Female Labor Force	0.46	0.05	0.070181	0.004
Working Age Residents Under 25	0.19	0.06	0.114681	0.007
Adult Enrollment	0.13	0.05	0.254453	0.013

* Note: Values correspond to Jobs-to-Labor-Force Mean= 0.87; Standard Deviation = 0.52

Other important factors in determining the predicted local working rate include: the proportion of residents who are black (standard deviation increase results in 1.9 percentage point decrease in local working); the proportion of adults 18-64 enrolled (standard deviation increase results in 1.3 percentage point increase in local working); the proportion of working-age residents under 25 (standard deviation increase results in 0.7 percentage point increase in local working); and average occupational level of employed residents (standard deviation increase results in 0.6 percentage point decrease). The strong effect of adult enrollment on local working is straightforward. Those enrolled, including college students working part-time, are much more likely to seek work and be employed near their neighborhood. The effect of average occupational level is predicted by Simpson and Granovetter, both of whom suggest that those at higher skill levels exhibit larger job search radii (Simpson, 1992; Granovetter, 1992). Unexpected, though, is the finding that the occupational level of nearby jobs does not significantly affect the local working rate.

The proportion black coefficient is particularly interesting. It suggests that blacks tend to attain work near their neighborhoods at a lower rate than nonblacks, even after controlling for the number and types of jobs near the neighborhood, their occupational level, and other factors. This relationship is also evidenced by the simple correlation between proportion black and local working of -0.3657 from Table III. The negative relationship between proportion black and local working cannot be explained by higher black unemployment, because, when substituting w, the proportion of employed residents working within two miles of the neighborhood, for w, proportion black still comes in negative, with an only somewhat smaller magnitude on the coefficient. (Results are not shown here.) The correlation between proportion black and the proportion of employed residents working locally is -0.3018.

These findings suggest that one of Kain's original premises, that blacks tend to work more in their own neighborhoods due to racial segregation, was no longer true by 1990 (Kain, 1968). Blacks may now tend to work farther from their neighborhoods due to relatively better opportunities elsewhere. The environment for black-owned small businesses in many black neighborhoods has certainly deteriorated over the last thirty years, so there are likely to be few employment opportunities at black-owned establishments, which disproportionately hire blacks (Bates, 1993). Perhaps more importantly, better opportunities have developed outside of black neighborhoods, with gains in employment at public sector and large employers, who tend to be located in the central business district and other areas and not near black neighborhoods. At the same time, blacks may find it hard to penetrate the hiring networks utilized by existing, non-black-owned, smaller firms near their neighborhoods, and these firms may be more prone to racial discrimination and less affirmative in their minority hiring than larger firms in other areas. This is also counter to the conclusions of Kain and others, who suggested that firms located in black neighborhoods were less likely to discriminate. Much has changed in these neighborhoods since the 1960s.

The effect of the proportion of working-age adults under age 25 is consistent with theory and the empirical literature. Youth are expected to work more in nearby jobs for several reasons. They may often seek part-time work because commuting and search costs make finding work elsewhere prohibitively expensive. Youth are also less likely to have accumulated sufficient wealth to purchase and maintain an automobile and so may be less mobile. Wilson argues that youth depend more on neighborhood and social networks to find work, and O'Regan and Quigley have found that employment prospects of youth are tied to those of their parents (Wilson, 1987; O'Regan and Quigley, 1993). Hanson and Pratt have found that workers relying on informal hiring networks tend to work closer to their place of residence (Hanson and Pratt, 1992).

Comparing Estimations of Unemployment and Local Working: Additional Support for a Nonlinear Specification

Comparing the results in Tables X and XI to those found in estimating effects on unemployment (Tables III through VI) reveals that the signs of the jobs-to-labor-force terms are the same. If job access alone was the cause of the relationship between jobs-to-labor-force and unemployment, then the corresponding coefficients should have opposite signs in the results of the two estimations. This is because, as increasing jobs-to-labor-force causes more residents to attain local work, it should also result in lower unemployment. The functions are expected, in effect, to "mirror" each other. As Figure 6 shows, however, increasing the jobs-to-labor-force ratio first results in increasing predicted unemployment while also resulting in higher predicted local working. This suggests that some other phenomenon is contributing to the relationship between jobs-to-labor-force and unemployment. One possible explanation is that unemployment-averse households prefer low jobs-to-labor-force areas over those with modest ratios, as postulated in Chapter 6. This would result in lower unemployment rates at low jobs-to-labor-force levels. At the same time, a low jobs-to-labor-force ratio forces a larger proportion of residents to find work outside the nearby job catchment area, resulting in a lower local working rate.

The local working rate results presented here (Table X and XI) are consistent with my explanation of the relationship between nearby jobs and unemployment. The cubic specification is able to capture some of the preference of lower unemployment households for low jobs-to-labor-force areas. When comparing neighborhoods with very low jobs-to-labor- force ratios to those with modest ratios, one sees that residents in the former neighborhoods tend to have low unemployment and, because there are few jobs nearby, local working rates are very low. In

Figure 6. Comparison of Unemployment and Local Working Functions.

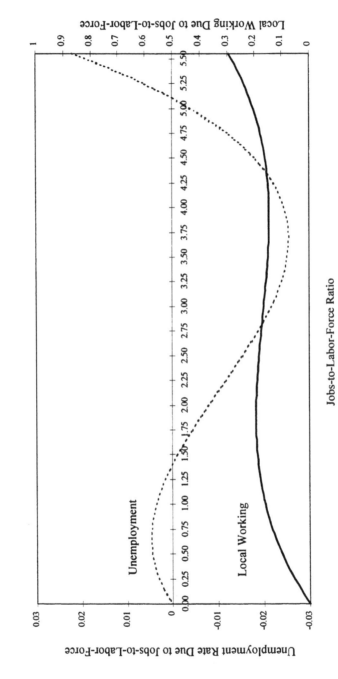

the latter neighborhoods, unemployment rates are somewhat higher because some highly employable households avoid neighborhoods with even modest jobs-to-labor-force ratios, but more residents are able to find work nearby.

This all suggests that, after discounting for the preference of unemployment-averse households for low jobs-to-labor-force areas, the impact of nearby jobs on neighborhood unemployment is greater than that estimated via the cubic specification. Thus, a standard deviation increase in jobs-to-labor-force may result in a decrease in predicted unemployment of more than 0.8 percentage points, after accounting for a potential preference for small jobs-to-labor-force levels among some households. After incorporating all job-related effects, the finding remains that job proximity and the nature of nearby jobs are, at best, only about as important as the average occupational level of neighborhood residents in determining unemployment. Perhaps most significant is the finding that race remains the single largest determinant of neighborhood unemployment—more important than either nearby labor demand or occupational level of residents.

CHAPTER 7

Implications for Neighborhood Economic Development and Employment Policy

FROM EMPIRICAL RESULTS TO POLICY ANALYSIS

The findings in Chapters 5 and 6, by themselves, are not sufficient for developing detailed recommendations for federal, state and local policies that seek to reduce neighborhood unemployment or promote local working in high-unemployment neighborhoods. Rather, the conclusions of the econometric analysis must be combined with what is known from the literature on urban employment patterns and neighborhood-level economic development policies.

I proceed carefully in stepping from interpretation of a particular analysis of employment patterns to policy prescriptions. I adopt the general approach of Lindblom, who advocates incrementalism in policy analysis, in which rigorous social science inquiries add to the overall knowledge of lay and professional knowledge (Lindblom, 1990). He recognizes the utility of social science in policy-making, especially in reducing the "impairment" caused by distorted information and perverse interests that enter into policy debates. But he dismisses the rigid notion of a scientific society, in which positivist analyses of human and firm behavior are sufficient for formulating policy. In Lindblom's framework, social scientists seek to add value to general social inquiry, which must include lay citizens and practitioners as well as academic scholars:

The self-guiding model rejects social science as an alternative to ordinary inquiry and sees it instead as an aid, refiner, extender, and sometimes tester of it, always a supplement, never broadly embarked on a program to displace or replace it. No less present in the self-guiding model than in the science model, social science pursues an adaptive strategy in the former instead of trying to set out on its own course as in the latter (Lindblom, 1990, pp. 216-217).

Any one type of policy action may affect a variety of the barriers to employment of local working identified in the econometric analysis. Therefore, I examine various policy areas and their potential to affect neighborhood unemployment or local working by reducing one or more barriers to employment. I do this by combining the findings from my research, which identify the most significant barriers, with information from the literature on the effectiveness and problems of particular approaches.

Table XII presents a simplified schematic of general policy implications stemming from the key findings of Chapter 6. For each key variable, the impacts of the variable on predicted unemployment and local working are described. Then, the general implications of these findings for neighborhood economic development and other relevant policies are described. A fuller analysis of policy implications is presented following Table XII, including some specific examples of implications for current federal policies.

I make no effort to assess the overall merits of any particular neighborhood economic development policy. Many of these policies seek to effect a broad spectrum of outcomes, including higher wages of local employed residents, the development of neighborhood businesses providing residents with goods and services, the reuse of older real estate properties, and the long-term development of community capital, leading to what is called "community building." Even if place-based economic development is not the most effective approach to reduce neighborhood unemployment, it may be necessary to maintain or develop a neighborhood's overall community capital. Otherwise, it is possible that, as the employment prospects of a subset of residents improve, they may leave the neighborhood, leaving it with high unemployment and low personal income once again. Improving the overall quality of community life is likely to help retain more employable residents in a neighborhood, thereby lowering

Table XII. Summary of Key Findings and Policy Implications

Key Variable	Effects on Unemployment and Local Working	Key Policy Implications
Jobs-to-Labor-Force Ratio (a ratio equal to the number of nearby low- and moderate-skilled jobs divided by the number of nearby low- and moderate-skilled residents in the labor force)	Unemployment: Significant, negative, modest impact: standard deviation increase results in approximately 0.8 percentage point decrease in unemployment. (mean unemployment = 7 percent) Local Working: Significant, positive, large impact: standard deviation increase results in 3.6 percentage point increase in local working. (mean local working = 16 percent)	Near-the-neighborhood job creation, by itself, will have only a modest impact on the neighborhood unemployment rate. Thus neighborhood economic development efforts that are not oriented to the neighborhood labor force should not be the principal method used to reduce neighborhood unemployment. At the same time, near-the-neighborhood job creation will boost neighborhood local working substantially, resulting in some welfare effects for local residents, including decreased commuting times and the potential for stronger employer-resident networks.

Table XII. Summary of Key Findings and Policy Implications (Continued)

Key Variable	Effects on Unemployment and Local Working	Key Policy Implications
Occupational Dissimilarity (an index measuring the dissimilarity between the occupations of nearby jobs and those of neighborhood residents)	Unemployment: Significant, positive, modest impact: standard deviation increase results in 0.7 percentage point increase in unemployment. Local Working: Significant, negative, modest impact: standard deviation increase results in 0.6 percentage point decrease in local working.	Near-the-neighborhood job creation efforts will have a larger impact on neighborhood unemployment, and on local working, if the mix of occupations of the jobs are similar to the occupational mix of neighborhood residents. Neighborhood economic development efforts that adopt "labor-force-based" strategies should be favored over those that do not. Similarly, labor-force-based development will increase the impact on local working, though to a relatively smaller extent.
Job Occupational Level (an index measuring the average occupational level of nearby jobs)	Unemployment: Significant, positive, modest impact: standard deviation increase results in 0.5 percentage point increase in unemployment. Local Working: Insignificant.	Near-the-neighborhood job creation efforts will have a larger impact on neighborhood unemployment if the average occupational level of the jobs is relatively low. Many unemployed residents will be relatively low-skilled. Therefore, labor-force-based development strategies in high-unemployment neighborhoods should focus on lower-skilled jobs.

Table XII. Summary of Key Findings and Policy Implications (Continued)

Key Variable	Effects on Unemployment and Local Working	Key Policy Implications
Proportion Black (proportion of neighborhood residents who are black)	Unemployment: Significant, positive, substantial impact: standard deviation increase results in 3.6 percentage point increase in unemployment. Local Working: Significant, negative, modest impact: standard deviation increase results in 1.9 percentage point decrease in local working.	Race effects may be due to employment discrimination, poor access to job networks, and unmeasured educational or skill differences. Policies to reduce neighborhood unemployment or increase local working should include race-based programs, including employment antidiscrimination and network-building. Specific recommendations include the enforcement of equal employment law, employment brokering, and minority business development.
Occupational Level (an index measuring the average occupational level of neighborhood residents)	Unemployment: Significant, negative, substantial impact: standard deviation increase results in 2.2 percentage point decrease in unemployment. Local Working: Significant, negative, modest impact: standard deviation increase results in 0.6 percentage point decrease in local working.	Education and job training policies are important to reducing unemployment in high unemployment neighborhoods. Economic development efforts need to be linked more closely to job training programs. If programs lack effectiveness but show potential for improvement, investments in improving these programs are warranted. These programs should be targeted to residents of high-unemployment neighborhoods.

unemployment and increasing local working. In addition, increasing local working may result in better local job networks in the long run, eventually leading to lower unemployment.

I do not evaluate the potential for such longer-term results here. Rather, I explore only the more direct potential of policies to achieve one or both of two employment status objectives: 1) decreased neighborhood unemployment; or 2) increased neighborhood local working. I first examine the implications of my findings for place-based neighborhood economic development policy and practice. I then consider implications of my findings for employment discrimination and job training and placement policies.

THE MEANING OF RACE: EDUCATION, DISCRIMINATION AND ACCESS TO JOB NETWORKS

The bridge from my econometric analysis to policy prescription is complicated by the fact that the importance of race (black vs. nonblack), in particular, may be representing several employment barriers, including employment discrimination, lack of job networks, and inferior educational quality. Blacks continue to be victims of employment discrimination (Turner et al., 1991, Kirschenman and Neckerman, 1991). Wilson argues that the increasing concentration of poverty in, and the exodus of middle-class blacks from, black neighborhoods have left many blacks with few personal connections to working adults (Wilson, 1987). The social isolation of many highly segregated, lower-income black neighborhoods leads to a disconnection from job networks, especially for lower-skilled residents. Massey and Denton argue that the residential hypersegregation of blacks leaves blacks with fewer contacts in the job market (Massey and Denton, 1993).

Orfield argues that the educational segregation and the inferior schooling of blacks and central-city students have led to large differentials between black and white student achievement as well as between the achievement of central-city and suburban students (Orfield, 1992; Orfield, 1990). He reports that, of 68 Chicago area firms responding to a Chicago Tribune survey in 1988, not one gave City of Chicago public schools, which serve a predominantly minority student body, an "A" or a "B" grade (Orfield, 1990). Twenty-five percent of the firms gave the schools a failing grade, with the average grade being a low "D". Two-thirds of the respondents believed schools were getting worse, and none said that the schools adequately prepared students for entry-level work.

Given the magnitude of the effect of race on unemployment in the analysis in Chapter 6 and the literature on the effects of education, networks and discrimination on black employment prospects, it is very possible that all three factors are significant contributors to the effect of race on unemployment. A case might be made that inequality in educational quality is likely to account for a larger proportion of the race effect than either employment discrimination or poor access to networks. The analysis in Chapter 6 may not fully control for differences in basic skills and educational achievement. I use three variables that serve to control for skills differences among neighborhoods: average resident occupational level, teenage enrollment in school, and adult enrollment in school. While occupational level should serve as a relatively strong control for substantial variations in educational achievement among moderate- and high-skilled residents, it cannot control adequately for modest variations in educational achievement, especially among lower-skilled residents where unemployment levels are the greatest. Those seeking low-skilled occupations may vary significantly in their educational achievement and basic skills. For example, average resident occupational level may not adequately capture differences between unemployed workers in one neighborhood with an average sixth-grade reading level and those in another with an average ninth-grade reading level. The addition of teenage and adult enrollment rates may also not fully capture such differences.

Despite the difficulty of fully controlling for educational achievement and skills, the fact that the proportion of neighborhood residents who are Hispanic does not significantly affect neighborhood unemployment suggests that educational differences are unlikely to explain the bulk of the importance of race. Many Hispanics also attend poorly performing city schools or speak little English. At the same time, Hispanics in Chicago do not suffer from the same substantial barriers to employment that blacks face. Holzer (1996) makes a similar points in describing the higher hiring rates of lower-skilled Hispanics vis-à-vis blacks.

IMPLICATIONS FOR PLACE-BASED POLICY AIMED AT REDUCING NEIGHBORHOOD UNEMPLOYMENT AND INCREASING LOCAL WORKING

General Implications: Linking Development to Residents

The primary economic development objective of interest here is the reduction of neighborhood unemployment. The results of Chapter 6 show that the effect of nearby jobs on neighborhood unemployment depends not only on the ratio of nearby jobs to nearby labor force, but also on the match of the occupations of the nearby jobs to the occupations of neighborhood residents and on the occupational levels of the nearby jobs. Combined, these three nearby labor demand characteristics can have a substantial effect on neighborhood unemployment. The relative magnitude of nearby jobs-to-labor-force, by itself, does not have a large effect on neighborhood unemployment, so any policy prescription for neighborhood economic development as a method of reducing neighborhood unemployment should consider the nature as well as the magnitude of nearby jobs.

The secondary economic development objective of interest here is increasing the neighborhood local working rate. Increasing the number of nearby jobs—and thus the jobs-to-labor-force ratio—will have a large, positive impact on the local working rate. Moreover, the greater the match between the occupations of nearby jobs and the occupations of residents, the greater will be the local working rate.

If the primary objective is to reduce neighborhood unemployment, job creation and retention efforts should favor occupations similar to those of residents and, in high-unemployment neighborhoods, efforts should be geared toward lower-skilled jobs. Other types of neighborhood economic development are not likely to reduce neighborhood unemployment very much. While lower-skilled jobs are likely to pay less than moderate-skilled ones, the moderate-skilled jobs may be accessible only to a small fraction of the unemployed—those who are more likely to find employment without development intervention. On the other hand, an increase in higher-wage, moderate-skilled jobs near the neighborhood may still yield benefits to local working, may raise the wages of appropriately skilled residents, and may result in other economic development benefits such as increased spending in the neighborhood by workers in local jobs. But to reduce unemployment, skill levels of the unemployed need to be brought up and racial barriers reduced before nearby jobs in moderate-skilled occupations are likely to reduce unemployment substantially.

Economic development programs that increase a neighborhood's jobs-to-labor-force ratio appear to be a promising approach to increasing local working. Relatively small increases in nearby jobs can significantly affect local working. Successfully matching created jobs to the occupations of residents will serve to increase local working even further. But the magnitude of the nearby jobs-to-labor-force ratio is by far the strongest determinant of local working.

The empirical work here confirms the common-sense notion that neighborhood job creation will lead to larger decreases in unemployment the more the new jobs are matched to the occupations of local residents. Ranney and Betancur refer to the practice of neighborhood economic development that is oriented around the occupations of residents as "labor-force-based development" (Ranney and Betancur, 1992). They argue that the traditional division between economic development practitioners and job training and placement specialists is a barrier to effective labor-force-based development. Economic developers often do not consider who will work in the firms they assist, while job trainers are constrained by the myriad of training program regulations with which they have to comply. Economic developers tend to view themselves as assisting the growth of firms while job training and placement agencies are viewed as social service providers.

Ranney and Betancur prescribe a more comprehensive strategy to developing job opportunities for neighborhood residents that involves the following basic steps: 1) identifying the occupational mix of unemployed neighborhood residents; 2) identifying stable or growing industries which employ substantial numbers of employees in such occupations; and 3) focusing nearby economic development efforts on these industries to create new jobs or identifying areas that are accessible to the neighborhood via mass transit and that have substantial employment in these industries.

Programs linking economic development to job training and placement of targeted populations are unusual (Bartik, 1994). Bartik suggests that most economic developers see their role as advocating for improving the local business climate and that they fear imposing hiring requirements may worsen this climate. Similarly, Giloth argues that most economic development has been concerned with jobs but usually as a secondary effect of a focus on how regional economies invest in infrastructure, technology and education (Giloth, 1995). He calls for economic development focused on how to improve employment access for disadvantaged people directly and in the short run.

The aversion to targeting is especially common at the state and local level. While federal targeted economic development efforts have frequently been concerned with employment impacts on disadvantaged populations and residents of target areas (the Community Development Block Grant (CDBG) program and Empowerment Zones being the two clearest examples), state and local economic development officials have often viewed their mission as improving productivity or expanding aggregate employment (Bartik, 1994). Fitzgerald describes how state economic development staff can be reluctant to coordinate their efforts with training efforts for the disadvantaged (Fitzgerald, 1993). Moreover, state and local constituent politics make any sort of targeting difficult, and the monitoring required to ensure hiring of certain targeted groups often taxes the capacity of small economic development staffs.

Stillman finds a number of barriers to improved coordination among economic development and job training and placement efforts (Stillman, 1994). Some of these barriers include:

Varying priorities
The priority of job training and placement professionals is finding jobs for the unemployed. While they typically understand that it is in their interest to view employers as clients, their primary client base remains the unemployed. At the same time, economic development officials see private firms and investors as their primary clients, and they often see the poor as a drag on their efforts to induce firm location, expansion or retention.

People vs. place
Job training and placement professionals tend to focus on the job needs of individuals. Economic development programs and personnel, on the other hand, often see the improvement of places or communities as the focus of their efforts.

Governmental fragmentation within metropolitan areas
The hundreds of distinct local governments in many metropolitan areas can present barriers to the coordination of economic development and job training and placement efforts. It is difficult enough to coordinate these two types of efforts within a single municipal government; coordination across urban-suburban and intersuburban boundaries poses even greater barriers.

To reduce some of these barriers, Stillman recommends cross training economic development and job training and placement staff, so those in the two different arenas learn and better understand the other's environment, motivations, resources and constraints (Stillman, 1994). He also recommends improving the understanding of local economies and labor markets among both types of agencies, and providing more flexible funding for creative and more efficient coordination of the two policy efforts.

In order to increase the effectiveness of neighborhood economic development efforts in reducing neighborhood unemployment, policy makers at all levels of government should support labor-force-based development strategies. That is, guidelines for firm incentives, subsidies and other economic development spending should favor those activities that adopt labor-force-based development approaches. Federal funding of local economic development should encourage such practice. Bartik has suggested that the federal government support demonstration programs that may test the potential for linking federal economic development and job-training programs (Bartik, 1994). This could be implemented at the neighborhood level, for example, by giving incentives to government and organizations funded by the CDBG program to form close alliances with those funded under federal job training programs. In fact, recent regulatory changes in the CDBG program permit funds to be used not only for job creation purposes, but also for job training and placement efforts tied to funded job creation efforts (Mayer, 1995).

One way that local governments have frequently attempted to tie economic development to job training and placement efforts is through first-source hiring agreements, in which firms receiving economic development assistance are asked to consider and give preference to job applicants from certain sources, usually either government or government-certified job training and placement organizations. These sorts of programs, for example, may encourage firms to consider more black job candidates from high-unemployment neighborhoods. Linking these programs with job training programs should enable these efforts to add value and provide an incentive for firms to take first-source agreements seriously. Mier suggests such efforts may be a good way to break into the informal hiring networks of employers that are often a formidable barrier to minority employment (Mier, 1993).

Stillman cites programs in St. Louis and Portland as examples of successful first-source efforts (Stillman, 1994). The St. Louis First-Source program requires local employers that receive public funds in the form of loans, tax abatements, incentives or other subsidies to reserve 60 percent of entry-level jobs openings for low-income applicants. The City of St. Louis certifies eligible job applicants, maintains a database of them, screens them, provides three candidates for each opening and monitors overall compliance. The program has averaged over 200 placements a year in recent years.

Portland's JobNet program is targeted at residents of a predominantly black, high-unemployment neighborhood in Portland (Stillman, 1994). JobNet staff work with local firms to develop a local hiring plan. Through a network of local job training organizations and community colleges, JobNet recruits, screens and trains area residents. A consortium of nine companies and five colleges has developed a standard curriculum for entry-level job training that can be customized for different firms.

Implications for Empowerment Zones and Other Federal Urban Economic Development Policies

Perhaps the most prominent initiative of the Clinton Administration's urban policy, Empowerment Zones are the most conspicuous of recent federal place-based urban economic development policies. Empowerment Zones can be as large as 20 square miles, with those in the higher-density cities being somewhat smaller. Chicago's Empowerment Zone, for example, is approximately 14.6 square miles in area, making it generally consistent in size with the job catchment area (12.25 square miles) used in my analysis. One of the key incentives included in Empowerment Zone designation is a 20 percent wage tax credit, capped at $3,000, available to firms located in and employing residents of a Zone. Other incentives include access to tax-exempt bond financing, which significantly reduces the interest rate on business financing, and an acceleration of depreciation on investments in property in the Zone of up to $37,500. The latter two incentives are conditional upon 35 percent of firm employees being residents of the Zone. Zones are also eligible for as much as $100 million of Social Service Block Grant funds that can be used for a wide variety of economic and labor force development uses (United States Department of Housing and Urban Development, 1994).

The Empowerment Zone employer tax credit is essentially a wage subsidy. It is similar in structure to the Targeted Jobs Tax Credit, a federal program which gives employers a tax credit for hiring economically disadvantaged individuals, except that the criterion for eligible employees is residential location and not income. The Targeted Jobs Tax Credit has been criticized for not expanding employment, so that employers collect credits for employees whom they would have hired without the subsidy (Dewar and Scheie, 1994). But some wage subsidies appear to induce firms to hire more employees. More than one-half of the firms assisted through the Minnesota Emergency Employment Development (MEED) program, which is no longer in existence, claimed they would not have expanded employment without the program (Dewar and Scheie, 1994; Rode, 1988).

The federal On-the-Job-Training program is essentially a wage subsidy paid during a training period of up to 6 months. There are reports that firms prefer to avoid the work involved in applying for the subsidy and that they feel that the subsidy should not influence the hiring process (Greater North-Pulaski Development Corporation, 1995). That is, they utilize the program only if it does not actually influence their hiring decision. Such attitudes may signal the stigma that can be associated with wage subsidies. Some research suggests that applicants who announce that they are eligible for a wage subsidy may be less likely to attain employment than if they do not reveal their eligibility (Burtless, 1985). The fact that only an estimated 3-4 percent of all employers eligible to use the Targeted Jobs Tax Credit take advantage of it could be due, in part, to such stigma (Eisner, 1989). Bartik suggests that customized screening programs can mitigate wage-subsidy stigma by providing employers with greater confidence in the quality of job applicants (Bartik, 1994). This problem may also be less likely if the subsidy runs to the place of residence, as in the case of the

Empowerment Zone, because employers will be able to discern residence without knowing of the subsidy.

The structure of the Empowerment Zone tax credit suggests that, to the extent that it increases the number of jobs in the Zone, it should result in a direct increase in local working. Moreover, because the tax credit is restricted to residents of the Zone, existing employees already living in the zone may be in a position to request higher wages, i.e., they may seek to extract some of the tax credit. If they left the firm, the firm may have to replace them with workers living outside the Zone, who would not be eligible for the credit. This is particularly plausible in the case of more skilled Zone residents, because it may be relatively difficult for firms to find higher-skilled workers living in the Zone.

It remains unclear, however, how much such credits will affect Zone unemployment. Firms may merely hire Zone residents who are currently employed outside the Zone. The effect on job growth will be larger if any new jobs are in occupations similar to those of unemployed Zone residents.

The investment subsidies of the Empowerment Zone program (accelerated depreciation and tax-exempt bond financing) may lead to increased employment as firms expand or move into the zone. These subsidies, themselves, are not likely to reduce Zone unemployment very much, although they are likely to increase local working if the investments lead to expansions of overall activity that in turn result in increases in Zone jobs. There is a possibility that a subsidy for capital may actually encourage the substitution of capital for labor, thereby resulting in a reduction in Zone jobs, even those held by nonresidents. The net jobs effect will depend on whether the output effect of the subsidy on jobs exceeds the substitution effect (Papke, 1994).

In addition to subsidies to firms located in Empowerment Zones, Zones are eligible for as much as $100 million in Social Service Block Grant funds for a wide variety of economic and labor force development uses. The guidelines for such projects are very broad, with a mandate for neighborhood-level, community-based planning. Eligible uses include business development activities such as business incubators, commercial real estate development and job-training activities. While the focus on resident participation in the Zone planning process may encourage labor-force-based development strategies, there is no assurance that such strategies will emerge. Job training efforts that can be tailored to the needs of a particular community would be consistent with the findings here on the importance of skills to neighborhood employment status.

In addition to direct and tax expenditures in the Empowerment Zone and Enterprise Communities program, other federal spending and regulatory initiatives seek to spur economic development in and around high-unemployment, lower-income neighborhoods, in part to improve employment prospects for residents of these neighborhoods. These programs are located in the Department of Housing and Urban Development, the Economic Development Administration (EDA), the Small Business Administration (SBA), the Treasury Department, and the Environmental Protection Agency (EPA).

Since the mid 1970s, the federal CDBG program has provided funds to local government for a wide variety of purposes, including housing development, targeted economic development, and general community betterment. Funded at approximately $4.8 billion for fiscal year 1995, CDBG provides funds to large cities directly and to small cities via state government. Job creation was identified as the primary purpose for only about six percent of CDBG funds for fiscal year 1991, but another 37 percent of funds were used for "area benefit, limited clientele, and economic development" projects, which can contain economic development projects (United States Department of Housing and Urban Development, 1995a). Housing is the largest single use of CDBG funds. CDBG regulations currently require that, for projects qualifying on job creation grounds, 51 percent of created jobs go to low- and moderate-income persons. If a state or local government chooses to use CDBG funds to spur jobs for residents of a particular neighborhood or set of neighborhoods, then labor-force-based economic development, described in section C.1., should be adopted.

The EDA's Competitive Communities Initiative seeks to support the development of high-growth, globally competitive businesses in distressed communities (United States Department of Housing and Urban Development, 1995b). Unfortunately, many such firms may be only small employers of lower-skilled labor, so that the occupational match between such jobs and distressed neighborhoods may be quite small. Unless special, intensive training efforts accompany such an effort, neighborhood unemployment is unlikely to be reduced significantly. Local working rates, however, can be increased to some degree through such measures, as higher-skilled, employed residents find work with the new employers. The mismatch between residents and jobs, however, will dampen the impact.

The SBA's One Stop Capital Shop initiative seeks to provide technical assistance with obtaining financing to firms in targeted, distressed urban neighborhoods, especially those in Empowerment Zones (United States Department of Housing and Urban Development, 1995b). The One Stop Capital Shop's primary purpose is business development and not necessarily job creation. Moreover, it is likely to have only a small impact on neighborhood unemployment, because it does not focus on businesses most likely to employ residents of target areas.

In the Treasury Department, the Community Development Financial Institutions (CDFI) Fund has a budget for fiscal year 1995 of $50 million for investments in and grants to banks and specialized community development lending entities targeting distressed urban and rural areas (United States Department of Housing and Urban Development, 1995b). Funded entities will provide capital and credit to both housing and economic development projects. The program was originally envisioned as a much larger program and was authorized at almost $400 million over four years. While capital investments in high unemployment areas are unlikely to reduce unemployment substantially, they are likely to increase local working and, if combined with attention to resident skills and occupations, they can have a more significant impact on unemployment. An example of a community development financial institution that combines business development in targeted areas with efforts to improve the job skills of residents is the Shorebank Corporation's Austin Initiative on the west side of Chicago. The bank provides development financing to firms and attempts to marry it with job training and placement efforts of residents in the same neighborhood area. Because many of the entities that will obtain funds from the CDFI Fund have missions of reducing unemployment and serving lower-income residents of their target areas, the Fund has more promise of making progress toward this goal than does the more general SBA One Stop Capital Shop.

The EPA administers the Brownfields Redevelopment Initiative, which aims to return vacant and potentially polluted urban properties to productive use (United States Department of Housing and Urban Development, 1995b). The agency is working with state and local governments and the private sector to clarify liability issues, streamline review and decision procedures, and develop cleanup methods. Many of these sites are located in or near high-unemployment urban neighborhoods, and the creation of jobs in these areas is a clear goal of the program. The agency is providing up to $200,000 for each of 50 local redevelopment projects and is considering various regulatory changes to facilitate reuse of contaminated land. Because such sites are often very suitable for light manufacturing and assembly, warehousing and other industrial uses, this initiative may offer potential for the development of substantial numbers of lower-skilled jobs near high-unemployment neighborhoods. While the impact of brownfield redevelopment, by itself, on neighborhood unemployment might not be very large, it should be larger than efforts less likely to create lower-skilled jobs. The impact on local working in nearby neighborhoods is expected to be substantial.

Overall, my findings suggest that, even if they are effective at spurring nearby business development, the impact of most federal neighborhood economic development efforts will, without connecting to targeted job training and placement programs, have only modest impacts on neighborhood unemployment. At the same time, such policies are likely to have much larger impacts on local working. The literature on enterprise zones reviewed in Chapter 2 suggests that place-based policies can alter the intrametropolitan or neighborhood location of businesses and jobs, but the cost-effectiveness of such efforts remains unclear. Notwithstanding these ongoing concerns, the findings suggest that targeted neighborhood economic development efforts can be more effective at reducing neighborhood unemployment and increasing local working if they focus on industries and companies that employ workers in occupations similar to those of neighborhood residents and if they emphasize the creation of lower-skilled jobs. Only after fundamental gains are made in the education and training of residents of high-unemployment areas will the creation of nearby, moderate-skilled jobs be likely to have a substantial impact on neighborhood unemployment.

IMPLICATIONS FOR RACE-BASED POLICIES TO REDUCE BLACK UNEMPLOYMENT

The finding that race is a major determinant of neighborhood unemployment has implications for three areas of race-based employment and economic development policy: employment discrimination law, employment brokering, and minority business development. First, given the importance of race in determining neighborhood unemployment and the persistence of employment discrimination, any attack on high levels of urban unemployment should include aggressive antidiscrimination policies as a cornerstone. Second, as an alternative to antidiscrimination laws, employment brokering and placement organizations can work to educate employers about the potential of black workers and discourage discriminatory behavior. Helping firms find qualified black job applicants can be an important step to changing the mindset of a discriminatory employer. Third, because minority businesses tend to employ greater percentages of minority employees, minority business development strategies can provide a complement to antidiscrimination efforts. Affirmative action contracting policies, although now under increasing pressure from federal courts, are an important component of such policies.

Equal Employment Opportunity Enforcement

Despite the impact of discrimination on black employment prospects, mention of efforts to combat employment discrimination is conspicuously absent from the Clinton Administration's urban policy statement (United States Department of Housing and Urban Development, 1995b). One problem is that antidiscrimination efforts can be costly and politically difficult to implement, especially when addressing discrimination among small firms. Public officials have difficulty addressing employment discrimination involving a politically powerful small business constituency. By way of contrast, the Clinton Administration continues to advocate aggressive enforcement of the Fair Housing Act as a tool to improve the employment opportunities of urban minorities (United States Department of Housing and Urban Development, 1995b). Antidiscrimination policy in employment, unlike in housing, has not come to be seen as a key urban policy issue or as a principal means to reduce unemployment among urban blacks.

The key federal laws covering employment discrimination are the Equal Pay Act and Title VII of the 1964 Civil Rights Act. Employment discrimination laws are enforced at federal, state and local levels with the federal government being the primary enforcer. The Equal Employment Opportunity Commission (EEOC) is the primary federal agency in this arena. While the concern here is with discrimination in hiring, as opposed to discrimination in compensation or promotions, relatively little of the EEOC's activity concerns hiring, amounting to less than 7 percent of charges filed in 1988 (Bendick, et al., 1994). This low level of activity may be partly due to the lack of information available to the job applicant being discriminated against. She may be told that a job has already been filled or that another applicant has been hired who is more qualified without having access to any information to corroborate or deny the employer's response. Therefore, discrimination in hiring cannot be enforced merely by responding to complaints of job applicants. EEOC reports on employment patterns must be examined proactively by the agency itself and other relevant agencies.

Bendick et al. argue that, during the 1980s, the enforcement activities of the EEOC suffered due to inadequate resources and leadership that was ambivalent about the agency's mission (Bendick, et al., 1994). Between fiscal year 1982 and fiscal year 1992, the agency's backlog of cases increased by almost 60 percent, from 33,417 to 52,856. Meanwhile, the portion of complaints dismissed by the EEOC for "no cause" rose from 29 percent in 1981 to 59 percent in 1986.

O'Connell cites U.S. General Accounting Office statistics that show that only about 3 percent of filed complaints resulted in findings of discriminatory behavior, termed cause findings, in 1987 (O'Connell, 1991). He examines a sample of discrimination complaints of a regional office of a state fair employment agency and attributes low cause findings to three factors: 1) conservative standards of proof associated with agency processes; 2) bureaucratic and legal constraints that investigators face; and 3) conservative, informal rules developed by investigators to cope with their constraints. Investigators relied heavily on unambiguous types of evidence, such as an employer expressing discriminatory intent or not following basic hiring routines. Moreover, while investigators were given performance goals based on how many filed cases they process, they were not given additional credit for positive cause findings, even though cause findings took more than twice as long to process and required the interviewing of 3.5 times as many witnesses, on average. Finally, complainant mistakes tend to result in no-cause findings, so that 73 percent of no-cause cases contained complainant mistakes. O'Connell makes a number of recommendations to improve EEOC and state employment discrimination enforcement operations, including: 1) mandate official employer routines and more thorough record keeping; 2) allow investigators to broaden investigation beyond the specifics of complainants' charges, expanding the ability to use occupational underrepresentation as evidence; and 3) adjusting investigators' performance goals to reflect the additional work involved in cause findings (O'Connell, 1991).

Clark makes a number of more fundamental recommendations to increase the power and authority of the EEOC in order to reduce employment discrimination (Clark, 1989). Clark's recommendations include:

1) The EEOC should be given cease and desist authority similar to that held by the National Labor Relations Board.
2) EEOC rules should be more definitive and less subject to regular revision by the courts. The agency should be given principal authority to interpret Title VII and the courts should be instructed to review EEOC interpretations only where the agency's interpretation appears to be in flagrant violation of the statute.
3) Litigation and enforcement of employment opportunity cases should be consolidated in the EEOC. (They are currently shared with the United States Departments of Labor and Justice.)
4) The EEOC's ability to bring private suits should be strengthened.

Without an active equal employment opportunity enforcement program, urban neighborhood unemployment is likely to remain a major problem in black neighborhoods.

Employment Brokering and Employer Education Aimed at Reducing Discrimination

While the aggressive enforcement of equal employment opportunity laws is needed to reduce neighborhood unemployment in black neighborhoods, excessive pursuit of discrimination enforcement may not be cost-effective, especially when dealing with large numbers of small firms. Rigorous investigation and enforcement aimed at a substantial portion of smaller firms could be expensive and, perhaps, politically unfeasible. Currently, firms with fewer than 100 employees are not required to report employment breakdowns by race to the Equal

Employment Opportunity Commission. Moreover, firms can mask discriminatory behavior by utilizing screening agents, such as unscrupulous or unwitting job placement organizations, or they may turn to temporary labor to avoid culpability.

Another method for combating discrimination, particularly among smaller firms, is the use of employment brokering and placement programs. Employment brokers are nonprofit organizations that seek to connect unemployed persons to job opportunities. Typically this involves targeting growing or stable industries with job opportunities that match those of the target population, gaining the trust of the industry, delivering training to the population as needed, and assisting with other aspects of the job search process (Dewar and Scheie, 1995; Harrison, et al., 1994; Stillman, 1994). Effective employment brokers develop strong relationships with employers and can help identify qualified black job applicants.

One approach that can be used by employment brokers to entice employers to consider black applicants involves offering employers a targeted wage subsidy such as the Empowerment Zone employment tax credit. While the Empowerment Zone tax credit is not strictly race-based, most residents of most Empowerment Zone neighborhoods are minorities, and in many cases, predominantly black. If the tax credit encourages discriminatory employers to hire black workers and the resulting experiences of the employers are positive, the subsidy may work to mitigate prejudice and reduce discriminatory behavior. While the Empowerment Zone employment tax credit may, by itself, not be very promising, it may be effective at combating racial discrimination if tied to an employment brokering strategy.

To improve the effectiveness of employment brokering, screening for basic skills and work attitude can help discourage biased views. This approach to reducing discrimination may prove more cost-effective and productive in the long run than frequent antagonistic investigations and enforcement measures targeting small firms. But some highly visible amount of equal opportunity enforcement is needed to discourage discrimination. Thus, the two approaches can complement each other.

One way that the public sector can address employment discrimination is to raise general awareness of the problem and support local partnerships aimed at reducing job bias. Nonprofit fair housing organizations, for example, have received federal funds not only to help enforce federal fair housing law but also to educate realtors and lenders about how to serve minority home buyers better. In a similar way, federal support could be targeted to job training organizations and employment brokers who would encourage firms to hire minority, and particularly black, job applicants. Such support could be channeled to organizations focusing on residents of high-unemployment neighborhoods. This support could also serve to raise the general level of public awareness of employment discrimination, especially as it affects lower-skilled black populations.

Minority Business Development Policy

Bates finds that minority firms disproportionately employ minority workers (Bates, 1993). Thus, an appropriate complement to encouraging nonminority-owned firms to hire blacks is to encourage employment growth in black-owned businesses. Not only are black-owned firms more likely to hire black residents, they may be more willing to invest in the training of black workers and promote black workers. Bates argues that the development of strong, profitable black-owned businesses across the metropolitan area is the most effective way to improve the employment prospects of blacks in low-income, high-unemployment neighborhoods.

Unfortunately, the available evidence on the performance of minority business development programs is not very encouraging. Discrimination in access to credit, lower levels of household wealth, lack of business contacts, and educational inequalities pose serious barriers to business formation and development among blacks. Due in part to these obstacles, some minority business development programs have not been very successful, suffering from low business survival rates, high loan losses in loan programs, and other problems (Bates, 1995).

Affirmative action policy may be important to maintain or increase black employment by supporting the growth of black-owned firms. Recent Supreme Court decisions, however, have seriously undermined minority business contracting programs within government agencies. The prospect for policy supporting black-owned business development is not very promising if the federal government itself turns more to nonminority-owned suppliers.

If some of the obstacles faced by black business development programs can be reduced and federal support for minority contracting programs does not continue to diminish, the potential remains for black business development to become more effective at growing black-owned businesses. Effective black business development programs have the potential to reduce neighborhood unemployment in black neighborhoods significantly. This is not to suggest that such policies, by themselves, would be sufficient to address black neighborhood unemployment problems. Even if the political climate changes and minority business development programs become more effective, black-owned businesses are likely to employ only a small fraction of the black urban population in the foreseeable future. These efforts should constitute only one part of a more comprehensive set of race-based policies.

IMPLICATIONS FOR EDUCATION, JOB TRAINING AND PLACEMENT POLICIES

The findings of Chapters 5 and 6 clearly support efforts to raise the skill levels of residents of high-unemployment neighborhoods in order to reduce unemployment. In fact, the findings suggest that skill mismatch is a larger factor in urban unemployment than spatial mismatch. A standard deviation increase in resident occupational level results in a 2.2 percentage point decrease in unemployment, compared to a mean unemployment rate of 7 percent. Neighborhood economic development policies should be clearly linked to these efforts and should encourage the reduction of institutional barriers between labor force development and economic development programs.

While my analysis does not fully control for the quality and quantity of basic educational attainment, the findings on occupational level and race, combined with the literature pointing to educational inequalities by race, suggest that primary and secondary education is a key factor in the determination of neighborhood unemployment rates. Chicago public schools are predominantly minority, and schools in the highest-unemployment neighborhoods are comprised of predominantly black students. The poor performance of public schools is a major barrier to employment for residents of these neighborhoods. Policy recommendations in this arena involve major changes in state fiscal policy, school system governance and educational integration policy.

Recommendations on these issues are beyond the scope of this study. One federal initiative that should be supported and targeted to high-unemployment neighborhoods is the School-to-Work Opportunities Act, which supports partnerships of employers, schools and others developing new work-based learning programs, apprenticeships and internships (United States Department of Housing and Urban Development, 1995b). This effort addresses both skill needs as well as access to job networks.

Job training and neighborhood economic development efforts need to be combined with placement efforts that are race-conscious, i.e., that recognize the greater barriers to employment faced by blacks and that work to overcome employment discrimination. While economic development and labor force development policies and practice need to be integrated better, there is a danger here. Job placement agencies must guard against unwittingly serving illicit purposes. Employers may be looking to meet "best-effort" equal employment opportunity requirements imposed by government agencies, without intending to hire applicants from agencies, or may be looking to "outsource" their discriminatory practices by instructing placement agencies to provide applicants of only certain ethnic origins (Greater North-Pulaski Development Corporation, 1995).

Substantial evidence exists that job training and placement policies need substantial improvement. The effects of the primary federal job training effort for disadvantaged adults and out-of-school youths, the Job Training and Partnership Act (JTPA) Title IIA, have been significant but not large. According to the principal, long-term evaluation of the program, JTPA Title IIA produces a modest increase in the earnings of adults over a 30-month period (Bloom, et al., 1994). The average 30-month gain in earnings is 8 percent for men and 15 percent for women. No effect is found for out-of-school youths. For men, three-fifths of this increase is attributed to an increase in hours worked, as opposed to an increase in hourly wages. For women, three-fourths of this increase is attributed to an increase in hours worked. The study finds significant effects for those recommended for on-the-job training but not for those recommended for classroom training in occupational skills. Adult women, but not men, recommended for other services such as job search assistance also enjoy a significant gain in earnings.

One problem with JTPA practice is that service providers may seek to cream the most qualified from the pool of potential applicants to enable the providers to meet placement criteria more easily (Barnow, 1989; Wong, 1990; Fitzgerald and MacGregor, 1993). The JTPA itself eliminated special financial incentives that were provided to states to reward them for training and placing members of the most disadvantaged groups (Bendick, 1989). On the other hand, some JTPA providers complain that, in order to maintain a "mixed portfolio" and achieve reasonable success, they must not be forced to serve only the most disadvantaged populations (Harrison et al., 1994). Federal guidelines should require states to provide additional funding for those providers seeking to place more difficult populations, while other providers should not be overly constrained in their training and placement activities. Overly constraining regulations can inhibit the types of institutional reform needed in this arena.

Recent state policy changes and federal proposals promise to radically alter the provision of job training and placement services. Federal legislation calls for the block granting of numerous sources of federal job-training funds to states and the implementation of coordinated services. Illinois, like many other states, has already implemented "one-stop" employment and training centers which would combine at one intake site some of the different services previously delivered by JTPA providers, the Illinois Department of Employment Security, and local private industry councils. All those participating in federal- and state-funded job training and placement programs are processed through such centers and then provided with a voucher with which they purchase job training services from a certified provider. One-stop centers are located throughout the metropolitan area and operated by for-profit or non-profit organizations contracting with the State of Illinois. The intent of the voucher system is to introduce more competition into the current job training system, in which private industry councils let contracts to nonprofit job training organizations. The new system requires job training organizations, including both nonprofit and for-profit organizations, to win the business of individuals who receive training vouchers from one-stop centers.

It is clear that job training policies are extremely relevant to concerns about neighborhood unemployment rates. These efforts need to be rigorously evaluated from their inception, and they must be given adequate resources. Voucher systems for job training may provide some increased incentive for agencies to provide more effective services. But questions remain about the ability of low-skilled workers to make informed decisions about service providers, about the potential for poor coordination and communication among job training providers, and about whether agencies serving more difficult clients will be compensated at higher levels. Moreover, more must be known about successful methods for targeting and delivering job training to blacks and residents of high-unemployment neighborhoods.

FINAL CONSIDERATIONS FOR CONTINUING RESEARCH AND POLICY DEBATES

Combined with the available research on neighborhood economic development and employment patterns, the findings of this study suggest that neighborhood economic development, without concern for the types of jobs being created, will have only modest effects on neighborhood unemployment rates. To increase the magnitude of these effects, it is important to design neighborhood development efforts to match the occupational characteristics of neighborhood residents, particularly those in lower-skilled occupations.

At the same time, neighborhood economic development of most types can have very substantial effects on local working rates. Increased local working may be beneficial for working parents who want to work close to home, for youth and those without cars who have to undergo significant hardship in commuting to low-paying jobs, and for long-term community building. Besides lowering unemployment and raising local working, other potential impacts of neighborhood economic development include: higher wages of employed residents; increased participation in the labor force among neighborhood residents; reduced neighborhood crime; the continued use and occupation of commercial real estate; and increased political attention to the neighborhood by state and local government. Research on the potential for neighborhood economic development efforts to effect these outcomes is scarce and should be pursued. One reason for the paucity of research in this area is the lack of quality data at small geographic levels. Federal census data do not track most of the outcomes of interest, and local data sources are often not designed for research purposes and so suffer from uniformity and quality problems. To conduct sound impact analyses of neighborhood economic development projects, it may be necessary to establish new, long-term data collection efforts in selected target and control areas.

If the principal objective of a set of policies is to reduce unemployment in high unemployment neighborhoods, race must be addressed as a driving factor in access to jobs. This is contrary to Wilson, who, while acknowledging race as important, argues for race-neutral policies that focus more on overall labor demand and upgrading of skills (Wilson, 1987). The magnitude of the race factor found here suggests that such approaches may be inadequate. Black neighborhoods suffer from much higher unemployment rates even after controlling for resident occupational level, age distribution, enrollment in school, and nearby jobs and occupational similarity to those jobs. An all-black neighborhood in the Chicago area, all other things being equal, is predicted to have an unemployment rate that is almost 11 percentage points higher than a similar all-white neighborhood.

Compounding the direct effect of race on unemployment is the fact that black neighborhoods also suffer from other conditions that lead to higher unemployment, including lower average occupational levels, higher levels of unemployment-prone youth, and lower nearby jobs-to-labor-force ratios. The result is that the *unadjusted* mean unemployment rate for neighborhoods that are all-black (more than 95 percent black) is 20 percent, while the mean unemployment rate among neighborhoods with no black residents is 4 percent, a 16 percentage point difference.

While the number of nearby jobs is the most important factor affecting local working, race is still important here. Even after controlling for other factors, black neighborhoods tend to have lower local working rates than nonblack neighborhoods. An all-black neighborhood in the Chicago area, all other things being equal is predicted to have a local working rate that is almost 6 percentage points lower than a similar all-white neighborhood.

Because black neighborhoods tend to have lower jobs-to-labor-force ratios, the *unadjusted* difference in local working rates between black and white neighborhoods is even larger. The mean local working rate for neighborhoods with no blacks is 17 percent, while the mean for all-black neighborhoods (more than 95 percent black) is only 7 percent, a 10 percentage point difference.

Continuing quantitative and qualitative research is needed to understand better the effects of spatial job distributions on employment rates and local working should neighborhood economic development, job training and employment brokering, antidiscrimination, and minority business development policies. This becomes increasingly important in the U.S. as low-skill inner-city residents are pushed into the job market by welfare reform. Neighborhood economic development should be integrated more with employment brokering and job training programs.

It is unlikely that any one single policy initiative will substantially reduce the problem of highly concentrated neighborhood unemployment. But better integration of policy efforts and more attention to race are in order. More experimentation, applied research, and creative civic discussion, especially regarding race, are needed to identify promising methods for integrating economic and labor force development policies and programs, for reducing employment discrimination against blacks, and for improving job training and placement services.

Bibliography

Barnekov, T., Boyle, R., and Rich, D. *Privatism and Urban Policy in Britain and the United States*, Oxford: Oxford University Press, 1989.

Barnow, B.S. Government Training as a Means of Reducing Unemployment. In: *Rethinking Employment Policy*, eds. D. L. Bawden and F. Skidmore, pp. 109-136. Washington: Urban Institute Press, 1989.

Bartik, T.J. *Who Benefits from State and Local Economic Development Policies?* Kalamazoo: Upjohn Institute, 1991.

Bartik, T.J. "The Effects of State and Local Taxes on Economic Development: A Review of the Recent Research." Economic Development Quarterly 6 (1992): 102-110.

Bartik, T.J. "Who Benefits from Local Job Growth: Migrants or the Original Residents?" *Regional Studies* 27(1993): 297-311.

Bartik, T.J. "What Should the Federal Government be Doing About Urban Economic Development?" Working Paper 94-25. Kalamazoo: Upjohn Institute, 1994.

Bates, T. "Why Do Minority Business Development Programs Generate So Little Minority Business Development?" *Economic Development Quarterly* 9 (1995): 3-14.

Bates, T. *Banking on Black Business*. Washington, D.C.: Joint Center for Political and Economic Studies, 1993.

Bauder, H. and Perle, E. "Job Accessibility and Spatial Mismatch for Detroit's Youth In Different Labor Market Segments." Presented at the 1995 conference of the American Collegiate Schools of Planning in Toronto.

Beaumont, E. "Enterprise Zones and Federalism." In: *Enterprise Zones: New Directions in Economic Development*, ed. R. E. Green, pp. 41-57. Newbury Park: Sage Publications, 1991.

Becker, G. S. *The Economics of Discrimination.* Chicago: University of Chicago Press, 1957.

Bendick, M., Jr. "Matching Workers and Job Opportunities: What Role for The Federal-State Employment Service?" In: *Rethinking Employment Policy,* eds. D. L. Bawden and F. Skidmore, pp. 81-108. Washington: Urban Institute Press, 1989.

Bendick, M.,Jr., Jackson, C.W. and Reinoso, V.A. "Measuring Employment Discrimination Through Controlled Experiments." *The Review of Black Political Economy* 23 (1994): 35-48.

Bloom, H.S., Orr, L.L., Cave, G., Bell, S.H., Doolittle, F. and Lin, W. *The National JTPA Study: Overview: Impacts, Benefits, and Costs of Title II-A.* Bethesda: Abt Associates, Inc., 1994.

Bluestone, B. and Harrison, B.: *The Deindustrialization of America: Plant Closings, Community Abandonment, and the Dismantling of Industry.* New York: Basic Books, 1982.

Bolton, R. "Place Prosperity Versus People Prosperity: An Old Issue With a New Angle." *Urban Studies* 29 (1992): 185-203.

Burtless, G. "Are Targeted Wage Subsidies Harmful? Evidence From a Voucher Experiment. *Industrial and Labor Relations Review* 39 (1985): 101-114.

Butler, S. "The Conceptual Evolution Of Enterprise Zones." In: *Enterprise Zones: New Directions in Economic Development,* ed. Roy E. Green, pp. 27-40. Newbury Park: Sage Publications, 1991.

Cain, G. "The Challenge of Segmented Labor Market Theories to Orthodox Theory: A Survey." *Journal of Economic Literature* 14 (1976): 1215-1257.

Carlson, V. and Theodore, N. "Employment Availability for Entry-Level Workers: An Examination of the Spatial Mismatch Hypothesis in Chicago." *Urban Geography* 18 (1997): 228-242.

Charney, A. "Intraurban Manufacturing Location Decisions and Local Tax Differentials." *Journal of Urban Economics* 14 (1983):184-205.

Clark, L.D. "Insuring Equal Opportunity In Employment Through Law." In: *Rethinking Employment Policy,* eds. D. L. Bawden and F. Skidmore, pp. 59-80. Washington: Urban Institute Press, 1989.

Cobb, J.C.: *The Selling of the South: The Southern Crusade for Industrial Development: 1936 -1990.* Urbana: University of Illinois Press, 1993.

Connecticut Development Authority. *Connecticut Small Business Reserve/Urbank,* brochure, Rocky Hill, Connecticut, 1994.

Cox, K. and Mair, A. "Locality and Community in the Politics of Local Economic Development." *Annals of the American Association of Geographers* 78 (1988): 118-127.

Dabney, D.Y. "Do Enterprise Zone Incentives Affect Business Location Decisions?" *Economic Development Quarterly* 5 (1991): 323-334.

Dewar, T. and Scheie, D.: *Job Opportunity Initiatives: Toward a Better Future for Low-Income Children and Youth.* Minneapolis: Rainbow Research, Inc., 1994

Eisenger, P.K. *The Rise of the Entrepreneurial State: State and Local Economic Development Policy in the United States.* Madison: University of Wisconsin Press, 1988.

Eisner, R. "Employer Approaches to Reducing Unemployment." In: *Rethinking Employment Policy*, eds. D. L. Bawden and F. Skidmore, pp. 59-80. Washington: Urban Institute Press, 1989.

Ellwood, D.T. "The Spatial Mismatch Hypothesis: Are There Teenage Jobs Missing in the Ghetto?" In: *The Black Youth Employment Crisis*, eds. R.B. Freeman and H. J. Holzer, pp. 147-187, Chicago: University of Chicago Press, 1986.

Erickson, R.A. "Enterprise Zones: Lessons From the State Government Experience." In: *Sources of Metropolitan Growth*, ed. E.S. Mills and J.F. McDonald, pp. 161-182, New Brunswick: Center for Urban Policy Research, 1992.

Fitzgerald, J. and McGregor, A. "Labor-Community Initiatives in Worker Training in the United States and the United Kingdom." *Economic Development Quarterly* 7 (1993): 172-182.

Fitzgerald, J. "Labor Force, Education And Work." In: *Theories of Local Economic Development: Perspectives from Across the Disciplines*, eds. R.D. Bingham and R. Mier, pp. 125-146. Newbury Park, Sage Publications, 1993.

Freeman, R.B. "Employment And Earnings of Disadvantaged Young Men in a Labor Shortage Economy." In: *The Urban Underclass*, eds. by C. Jencks and P. Peterson, pp. 103-121, Washington: D.C., Brookings Institution, 1991.

Giloth, R. "Social Investment in Jobs: Foundation Perspectives on Targeted Economic Development During The 1990s." *Economic Development Quarterly* 9 (1995): 279-289.

Goetz, E.G. "Type II Policy and Mandated Benefits in Economic Development." *Urban Affairs Quarterly* 26 (1990): 170-190.

Granovetter, M. "The Sociological and Economic Approaches to Labor Market Analysis: A Social Structural View." In: *The Sociology of Economic Life*, eds. M. Granovetter and R. Swedberg, pp. 233-263. Boulder: Westview Press, 1992.

Greater North-Pulaski Development Corporation. *Employment and Training Programs: An Evaluation from the Users' and Service Providers' Perspectives*. Chicago: Greater North-Pulaski Development Corporation, 1995.

Hansen, S.B. "Comparing Enterprise Zones to Other Economic Development Techniques." In: *Enterprise Zones: New Directions in Economic Development*, pp. 7-26. Newbury Park, Sage Publications, 1991.

Hanson, S. and Pratt, G. "Dynamic Dependencies: A Geographic Investigation of Local Labor Markets." *Economic Geography* 68 (1992): 373-405.

Harrison, B: *Education, Training and the Urban Ghetto*. Baltimore, Johns Hopkins University Press, 1972.

Harrison, B., Weiss, M., and Gant, J.: *Building Bridges: Community Development Corporations and the World of Employment Training*. New York, Ford Foundation, 1994.

Hill, D.: *Tribune 40: The First Forty Years of a Socialist Newspaper*. London, Quartet Books, 1977.

Hodge, C. L. *The Tennessee Valley Authority: A National Experience in Regionalism*. New York, Russell and Russell, 1968.

Holzer, H.J. "The Spatial Mismatch Hypothesis: What Has the Evidence Shown?" *Urban Studies* 28 (1991): 105-122.

Holzer, H.J. *What Employers Want: Job Prospects for Less-Educated Workers*. New York: Russell Sage Foundation, 1996

Holzer, H.J., Ihlanfeldt, K.R. and Sjoquist, D.L. "Work, Search, and Travel Among White and Black Youth." *Journal of Urban Economics* 35 (1994): 320-345.

Hughes, M.A. "Misspeaking Truth to Power: A Geographical Perspective on the 'Underclass' Fallacy." *Economic Geography* 65 (1989): 187-207.

Ihlanfeldt, K.R. *Job Accessibility and the School Enrollment of Teenagers*. Kalamazoo, Upjohn Institute, 1992.

Immergluck, D. *Focusing In: Indicators of Economic Change in Chicago's Neighborhoods*. Chicago: Woodstock Institute, 1994.

Indiana Department of Commerce. *Indiana Enterprise Zones: A Program Evaluation for 1989 and 1990*, 1992.

Jacobs, J. *The Death and Life of Great American Cities*. New York: Vintage Books, 1961.

Jencks, C. and Mayer, S.E. "Residential Segregation, Job Proximity and Black Job Opportunities." In: *Inner-City Poverty in the United States*, eds. L. E. Lynn, Jr. and M.G.H. McGeary, pp. 118-186. Washington, D.C.: National Academy Press, 1990.

Jencks, C. and Peterson, P., eds. *The Urban Underclass*, Washington, D.C., The Brookings Institution, 1991.

Johnson, J.H., Jr. and Oliver, M.L. "Structural Changes in the U.S. Economy and Black Male Joblessness: A Reassessment." In: *Urban Labor Markets and Job Opportunity*, eds. G. E. Peterson and W. Vroman, pp. 113-147. Washington, D.C.: Urban Institute Press, 1992.

Kain, J.F. "Housing Segregation, Negro Employment, and Metropolitan Decentralization." *The Quarterly Journal of Economics* 82 (1968): 175-197.

Kain, J.F. "The Spatial Mismatch Hypothesis, Three Decades Later." *Housing Policy Debate* 3 (1992): 371-460.

Kain, J.F. and Persky, J. "Alternatives to the Gilded Ghetto." *The Public Interest* 14 (1969): 74-87.

Kasarda, J.D. "Urban Industrial Transition and the Underclass." In: *The Ghetto Underclass: Social Science Perspectives*, ed. W. J. Wilson, pp. 43-64. Newbury Park, Sage Publications, 1993.

Kirschenman, J. and Neckerman, K. "We'd Love to Hire Them, But . . . : The Meaning of Race for Employers." In: *The Urban Underclass*, eds. by C. Jencks and P. Peterson, pp. 203-232. Washington, D.C., Brookings Institution, 1991.

Ladd, H.F. "Spatially Targeted Economic Development Strategies: Do They Work?" *Cityscape: A Journal of Policy Development and Research* 1 No. 1 (1994): 193-218.

Lemann, N. "The Myth of Community Development." *New York Times Magazine*, pp. 26-60, January 9, 1994.

Leonard, J.S. "Comments On: The Spatial Mismatch Hypothesis: Are There Teenage Jobs Missing in the Ghetto? In: *The Black Youth Employment Crisis*, eds. R. B. Freeman and H. J. Holzer, pp. 185-190. Chicago: University of Chicago Press, 1986.

Lindblom, C.E. *Inquiry and Change: The Troubled Attempt to Understand and Shape Society*. New Haven: Yale University Press, 1990.

Logan, J. and Molotch, H. *Urban Fortunes: The Political Economy of Place*. Berkeley, University of California Press, 1987.

Massey, D. and Denton, N. *American Apartheid*, Cambridge, Harvard University Press, 1993.

Mayer, C. J. "Does Location Matter?" *New England Economic Review*, May/June 1996, 26-40.

Mayer, N.S.: "HUD's First 30 Years: Big Steps Down a Longer Road." *Cityscape: A Journal of Policy Development and Research* 1 No. 3 (1995): 1-29.

McDonald, J.F. and Prather, P.J.: Suburban Employment Centers: the Case of Chicago. *Urban Studies* 31 (1994): 201-218.

Mier, R. "Community Development and Diversity." In: *Social Justice and Local Development Policy*, ed. R. Mier, pp. 182-200, Newbury Park, Sage Publications, 1993.

Mier, R. and Moe, K. "Decentralized Development: From Theory to Practice." In: *Harold Washington and the Neighborhoods: Progressive City Government in Chicago, 1983-1987*, eds. P. Clavel and W. Wiewel, pp. 64-99. New Brunswick, Rutgers University Press, 1991.

O'Connell, L. "Investigators At Work: How Bureaucratic and Legal Constraints Influence the Enforcement of Discrimination Law." *Public Administration Review* 51 (1991), 123-130.

O'Regan, K.M. and Quigley, J.M. "Labor Markets and Urban Youth." *Regional Science and Urban Economics* 21: 277-294, 1991.

O'Regan, K.M., Quigley, J.M.: Family Networks and Youth Access to Jobs. *Journal of Urban Economics* (1993), 34: 230-248.

Orfield, G. "Wasted Talent, Threatened Future: Metropolitan Chicago's Human Capital and Illinois Public Policy." In: *Creating Jobs, Creating Workers: Economic Development and Employment in Metropolitan Chicago*, ed. L. B. Joseph, pp. 129-160. Chicago, University of Chicago, 1990.

Orfield, G. "Urban Schooling And The Perpetuation of Job Inequality in Metropolitan Chicago." In: *Urban Labor Markets and Job Opportunity*, eds. G. E. Peterson and W. Vroman, pp. 161-199. Washington, D.C., Urban Institute Press, 1992.

Papke, L.E. "What Do We Know About Enterprise Zones." In: *Tax Policy and the Economy 7*, ed. J. Poterba, National Bureau of Economic Research, pp. 37-72. Cambridge, MIT Press, 1993.

Peterson, G.E. and Vroman, W., eds. *Urban Labor Markets and Job Opportunity*, Washington, D.C., Urban Institute Press, 1992.

Pierce, N.R. and Steinbach, C.F. *Enterprising Communities: Community-Based Development in America*. Washington, D.C. Council for Community-Based Development, 1990.

Porter, M. "The Competitive Advantage of the Inner City." *Harvard Business Review*, May/June, 1995.

Ranney, D. and Bentancur, J. J. "Labor-Force-Based Development: A Community-Oriented Approach to Targeting Job Training and Industrial Development." *Economic Development Quarterly* 6 (1992): 286-296.

Raphael, S. The Spatial Mismatch Hypothesis Of Black Youth Unemployment: Evidence From the San Francisco Bay Area. Unpublished manuscript. University of California at Berkeley, 1989.

Robinson, C.J. Municipal Approaches to Economic Development: Growth and Distribution Policy. *Journal of the American Planning Association 55* (1989): 283-295.

Rode, P. *MEED Means More Business Job Growth Through Minnesota Wage Subsidy Program*. Minneapolis, Jobs Now Coalition, 1988.

Rogers, C.L. "Job Search and Unemployment Duration: Implications for the Spatial Mismatch Hypothesis," *Journal of Urban Economics* 42 (1997): 109-132.

Rubin, B.M. and Wilder, M.G. "Urban Enterprise Zones: Employment Impacts and Fiscal Incentives. *Journal of the American Planning Association 55* (1989): 418-431.

Sanders. H.T. "Politics and Urban Public Facilities." In: *Perspectives on Urban Infrastructure*, ed. R. Hanson, pp. 4-66. Washington, D.C., National Academy Press, 1984.

Simpson, W.: *Urban Structure and the Labor Market: Worker Mobility, Commuting and Unemployment in Cities.* Oxford, UK., Clarendon Press, 1992.

Soot, S. and Sen, A.: A spatial employment and economic development model. *Papers in Regional Science* 70 (1992) 149-166.

Stillman, J.: *Making the Connections: Economic Development, Workforce Development and Urban Poverty.* New York, The Conservation Company, 1994.

Turner, M., Fix, M., and Struyk, R.: *Opportunities Denies, Opportunities Diminished: Discrimination in Hiring.* Washington, D.C., Urban Institute Press, 1991.

Turner, S. "Barriers to a Better Break: Employer Discrimination and Spatial Mismatch in Metropolitan Detroit." *Journal of Urban Affairs* 19 (1997): 123-141.

United States Bureau of the Census. *Journey to Work and Migration Statistics Branch: 1990 Census Transportation Planning Package Urban*

Element—Parts 1, 2 and 3 Technical Documentation for Summary Tape. Washington, D.C., August, 1993.

United States Department of Housing and Urban Development. *Guidebook for Community-Based Strategic Planning for Empowerment Zones and Enterprise Communities,* HUD-1442-CPD, 1994.

United States Department of Housing and Urban Development. *Consolidated Annual Report to Congress on Community Development Programs,* HUD-1513-CPD, 1995a.

United States Department of Housing and Urban Development. *Empowerment: A New Covenant with America's Communities: President Clinton's National Urban Policy Report.* Office of Policy Development and Research, 1995b.

United States Department of the Treasury, Community Development Financial Institutions Fund 12 Code of Federal Regulations Chapter XVIII et al., *Federal Register* Volume 60 Number 202 (October 19, 1995) pp. 54110-54141.

United States General Accounting Office. *Enterprise Zones: Lessons from the Maryland Experience,* GAO/PEMD-89-2, 1988.

Vidal, A.: Reintegrating Disadvantaged Communities into the Fabric of Urban Life: The Role of Community Development. *Housing Policy Debate* 6 (1995): 169-230.

Waddell, P. A Multinomial Logit Model of Race and Urban Structure. *Urban Geography* 12: 127-141, 1992.

Willis, K.G. "Estimating the Benefits of Job Creation from Local Investment Subsidies." *Urban Studies* 22 (1985): 163-177.

Wilson, W.J.: *The Truly Disadvantaged,* Chicago, University of Chicago Press, 1987.

Wilson, W. J. "Another Look at the Truly Disadvantaged." *Political Science Quarterly* 106 (1991-1992): 639-656.

Wilson, W.J. *When Work Disappears: The World of t he New Urban Poor.* New York: Knopf, 1996.

Winnick, L.: Place Prosperity Vs. People Prosperity: Welfare Considerations in the Geographic Redistribution of Economic Activity. In: *Essays in Urban Land Economics in Honor of the Sixty-Fifth Birthday of Leo Grebler,* pp. 273-283. Los Angeles, University of California at Los Angeles Real Estate Research Program, 1966.

Wolf, M. A. "Enterprise Zones: A Decade of Diversity. *Economic Development Quarterly:* 4 (1990): 3-14.

Wong, D. W. "Spatial Indices of Segregation." *Urban Studies* 30 (1993): 559-572.

Wong, K.K. "Toward More Effective Job Training in Metropolitan Chicago." In: *Creating Jobs, Creating Workers: Economic Development and Employment in Metropolitan Chicago*, ed. L. B. Joseph, pp. 129-160. Chicago, University of Chicago, 1990.

Wood, R.C. "People Versus Places: The Dream Will Never Die." *Economic Development Quarterly* 5 (1991): 99-103.

Zax, J. and Kain, J. F. "Moving to the Suburbs: Do Relocating Companies Leave Their Black Employees Behind? *Journal of Labor Economics* 14 (1996): 472 -504.

Index

Milton Keynes UK
Ingram Content Group UK Ltd.
UKHW031151141024
449569UK00024B/874